Springer Series on Naval Architecture, Marine Engineering, Shipbuilding and Shipping

Volume 21

The Naval Architecture, Marine Engineering, Shipbuilding and Shipping (NAMESS) series publishes state-of-art research and applications in the fields of design, construction, maintenance and operation of marine vessels and structures. The series publishes monographs, edited books, as well as selected PhD theses and conference proceedings focusing on all theoretical and technical aspects of naval architecture (including naval hydrodynamics, ship design, shipbuilding, shipyards, traditional and non-motorized vessels), marine engineering (including ship propulsion, electric power shipboard, ancillary machinery, marine engines and gas turbines, control systems, unmanned surface and underwater marine vehicles) and shipping (including transport logistics, route-planning as well as legislative and economical aspects).

The books of the series are submitted for indexing to Web of Science.

All books published in the series are submitted for consideration in Web of Science.

Fidaa Karkori

Dynamic Positioning Systems

Class Guidance for DP Operators

Springer

Fidaa Karkori
Southampton, UK

ISSN 2194-8445 ISSN 2194-8453 (electronic)
Springer Series on Naval Architecture, Marine Engineering, Shipbuilding and Shipping
ISBN 978-3-031-59172-3 ISBN 978-3-031-59173-0 (eBook)
https://doi.org/10.1007/978-3-031-59173-0

This Springer imprint is published by the registered company Springer Nature Switzerland AG
The registered company address is: Gewerbestrasse 11, 6330 Cham, Switzerland

Paper in this product is recyclable.

Preface

The application of dynamic positioning (DP) systems has been expanded significantly not only in the number of DP vessels, but also in the range of applications and the advancement of DP technologies. In order to address various needs of the industry, the majority of leading Classification Societies (Class) have developed Class specific guidance relating to the development, implementation and use of Dynamic Positioning Systems (DP). This book provides generalised guidance adapted from various leading Classes, to provide DP professionals with a readily available source of information and guidance relating to the operation and use of DP systems. The book provides optional notations and technical specifics that reflect current industry practice and DP technologies. Given it would be a mammoth undertaking in and of itself to cover every Class and notation contingency, the book focuses primarily on the American Bureau of Shipping (ABS) requirements for the design and testing of DP systems, as these are widely considered some of the most stringent and exacting requirements placed on the maritime industry. To that end, this book includes guidance relating to:

- Enhanced system (EHS) notations to recognise design features beyond current DPS-series notations and to provide flexibility to owners and operators.
- Station keeping performance (SKP) notations to recognise DP capability and to encourage robust design of the DP systems.
- Increased level of detail on the technical requirements to help less experienced users.
- Details of vessel type specifics to reflect industry need.
- Requirements for testing comparable with current industry practice.

The guidance covered within this book is applicable to systems that are installed onboard vessels, offshore installations and facilities. It is applicable to new build constructions.

Southampton, UK
2024

Fidaa Karkori

Introduction

The requirements contained in this book are for dynamic positioning systems for ships, mobile offshore drilling units, mobile offshore units and offshore support vessels. These requirements may be used in conjunction with Part 4 of the *ABS Rules for Building and Classing Marine Vessels* (Marine Vessel Rules), Parts 4, 6 and 7 of the *ABS Rules for Building and Classing Mobile Offshore Unit*s (MOU Rules) and the *ABS Rules for Survey After Construction* (Part 7) for vessels and offshore installations subject to the provisions of the ABS classification regime. For vessels and offshore installations subject to alternative classification regimes, refer to the specific Class Rules relating to those provisions. The requirements in the following sections are particularly relevant.

- *ABS Rules for Building and Classing Marine Vessels*
  - Internal Combustion Engines—Section 4-2-1
  - Gas Turbines—Section 4-2-3
  - Electric Motors and Motor Controllers—Section 4-8-3
  - Gears—Section 4-3-1
  - Propulsion Shafting—Section 4-3-2
  - Propellers—Section 4-3-3
  - Piping System—Chapter 4-6
  - Thrusters—Section 4-3-5
  - Control Equipment—Section 4-9-9

- *ABS Rules for Building and Classing Mobile Offshore Units*
  - Pumps and Piping Systems—Chapter 4-2
  - Electrical Installations—Chapter 4-3
  - Rules for Equipment and Machinery Certification—Part 6
  - Surveys After Construction—Chapter 7-2

- *ABS Rules for Survey After Construction*
  - Machinery Surveys—Chapter 7-6
  - Shipboard Automatic and Remote-control Systems—Chapter 7-8
  - Survey Requirements for Additional Systems and Services—Chapter 7-9

Classification Notation

The dynamic positioning systems are built and tested in compliance with the guidance outlined in this book and the relevant Class Rules may be assigned with different classification notations depending on the degree of redundancy built into the system as defined below. These notations are not a requirement for the classification of the vessel and are to be assigned only on the specific request of the Vessel Owner.

DPS-0 For vessels which are fitted with centralised manual position control and automatic heading control system to maintain the position and heading (Station Keeping) under the specified maximum environmental conditions.

DPS-1 For vessels which are fitted with a dynamic positioning system which is capable of automatically maintaining the position and heading (Station Keeping) of the vessel under specified maximum environmental conditions having a manual position control system.

DPS-2 For vessels which are fitted with a dynamic positioning system which is capable of automatically maintaining the position and heading (Station Keeping) of the vessel within a specified operating envelope under specified maximum environmental conditions during and following any single fault, excluding a loss of compartment or compartments.

DPS-3 For vessels which are fitted with a dynamic positioning system which is capable of automatically maintaining the position and heading (Station Keeping) of the vessel within a specified operating envelope under specified maximum environmental conditions during and following any single fault, including complete loss of a compartment due to fire or flood.

DPS-1, **DPS-2** and **DPS-3** classification notations are structured in line with the *IMO MSC.1/Circ.1580 "Guidelines for Vessels and Units with Dynamic Positioning (DP) Systems"*. **DPS-1**, **DPS-2** and **DPS-3** are in line with IMO equipment class of 1, 2 and 3, respectively.

At the Vessel Owner's request, the symbol + for the DPS-series notations may be assigned indicating additional requirements for station keeping capacity and failure modes for static components of the DP systems. At the Vessel Owner's request, an enhanced system notation (**EHS**), e.g., **EHS-E** and **EHS-P**, as a supplement for **DPS**-series notations, may be assigned. The notation provides the basis for the measurement to enhance critical components of the DP system, including power system, thruster system, control system and verification and validation of functionality. The main objective of the enhanced system notation is to improve reliability, operability and maintainability. Chapter 7 of this book provides details for the enhanced system notations. Also, at the Vessel Owner's request, a station keeping performance notation (**SKP**), as a supplement to the DPS-series notations may also be assigned to the vessel. The objective of the station keeping performance notation is to recognise the DP capability and to encourage robust design of DP systems. Chapter 8 of this book provides details for station keeping performance requirements. Chapter 10 of this book provides information about other optional notations that may apply to DP systems such as those for Integrated Software Quality Management (ISQM) and Software System Verification (SSV).

Definitions

The following definitions of symbols and terms are to be understood (in the absence of other specifications) where they appear throughout this book.

General

Dynamically positioned vessel (DP vessel): A unit or a vessel which automatically maintains its position and/or heading (fixed location, relative location or predetermined track) by means of thruster force.

Failure: An occurrence in a component or system that causes one or both of the following:

- Loss of component or system function; and/or
- Deterioration of functional capability to such an extent that the safety of the vessel, personnel or environmental protection is significantly reduced.

Failure Modes and Effects Analysis (FMEA): a systematic analysis of systems and subsystems to a level of detail that identifies all potential failure modes down to the appropriate subsystem level, and their consequences.

FMEA proving trials: the test programme for verifying the FMEA.

Industrial mission: Industrial mission is the primary operational role of the vessel, typically applicable to MODUs and project and construction vessels (e.g., pipe-lay/heavy-lift). (*Note*: *by definition the industrial mission for a logistic vessel is to support logistics*).

Logistics vessels: Offshore support vessels (OSV) and crew boats transporting, carrying out and facilitating the logistics competence of cargoes, goods, supplies, crews, loading/unloading, offshore exploration and production equipment, etc., across the maritime and offshore industries.

Specified maximum environmental conditions: The specified maximum environmental conditions are the specified wind speed, current and wave height under which the vessel is designed to carry out intended operations.

Specified operating envelope: The specified operating envelope is the area within which the vessel is required to stay in order to satisfactorily perform the intended operations under the specified maximum environmental conditions.

Worst-case failure (WCF): The identified single fault in the DP system resulted in maximum effect on DP capability as determined through the FMEA study. This worst-case failure is to be used in the consequence analysis.

Worst-case failure design intent (WCFDI): The specified minimum DP system capabilities to be maintained following the worst-case failure. The worst-case failure design intent is used as the basis of the design. This oftentimes refers to the simultaneous failure of multiple thrusters and/or generators.

Equipment and Dynamic Positioning System

Active component: Active components or systems requiring external power, such as generators, thrusters, switchboards, all related DP computer and communication networks, sensors, remote controlled valves, compensators, etc.

Static component: Static components are in particular: cables, pipes, manual valves, etc.

Dynamic positioning control station (DP control station): A workstation designated for DP operations, where necessary information sources, such as indicators, displays, alarm panels, control panels and internal communication systems are installed, including DP control and independent joystick control operator stations, required position reference systems' Human Machine Interface (HMI), manual thruster levers, mode change systems, thruster emergency stops and internal communications.

Dynamic positioning operation (DP operation): Use of the DP system to control at least two degrees of freedom in the horizontal plane automatically.

Dynamic positioning system (DP system): The complete installation necessary for dynamically positioning a vessel including, but not limited to, the following subsystems.

- Power system.
- Thruster system.
- DP control system.
- Independent joystick system (when required).

Power System: All components and systems necessary to supply the DP system with power, the power system includes, but not limited to:

- Prime movers with necessary auxiliary systems including piping, fuel, cooling, pre-lubrication and lubrication, hydraulic, pre-heating and pneumatic systems.
- Generators.
- Switchboards.
- Electrical distribution system (cabling and cable routing).

- Power management system if applicable.
- Power supplies, including uninterruptible power supplies (UPS).

Thruster system: All components and systems necessary to supply the DP system with thrust force and direction. The thruster system includes:

- Thrusters with drive units and necessary auxiliary systems including piping, cooling, hydraulic and lubrication systems.
- Main propellers and rudders if these are under the control of the DP system.
- Thruster control systems.
- Manual thruster controls.
- Associated cabling and cable routing.

DP control system: All control components and systems, hardware and software necessary to dynamically position the vessel. The DP control system consists of the following:

- Computer system/joystick system.
- Position and heading reference systems.
- DP sensor system.
- Control station and display system (operator panels).
- Associated cabling and cable routing.
- Networks.

Computer system: A system of one or more programmable electronic devices, associated software and hardware peripherals and interfaces. Microprocessors, Programmable Logic Controller (PLC), Distributed Control Systems (DCS) and PC or server-based computation systems are examples of computer-based systems.

Interface: A transfer point (includes physical and logical interfaces) at which information is exchanged. Examples of interfaces include input/output interface (for interconnection with sensors and actuators); and communications interface (to enable serial communications/networking with other computers or peripherals).

Peripheral: A device performing an auxiliary function in the system (e.g., printer, data storage device).

Joystick system: A system with centralised manual position control and manual or automatic heading control.

Position reference system: All hardware, software and sensors that supply information and/or corrections necessary to give position and heading references, including its power supply.

DP sensor system: A system comprising devices that measure vessel heading (such as gyro compasses), vessel motions (such as motion reference units) and wind speed and direction.

Consequence analyser: A software function which continuously verifies heading and position and issues an alarm if the vessel (in its current operating mode) in the current weather conditions is not able to keep the heading and position in the case that any of the predefined worst-case failures should occur.

Control mode: Method used for station keeping. Control mode for a DP system may be:

- DP control mode (automatic position and heading control).
- Manual position control mode (centralised manual position control with selectable automatic or manual heading control).
- Auto track mode, auto follow or follow target mode (considered as a variant of DP position control, with programmed movement of reference point).
- Manual thruster control mode (individual control of thrust (pitch or speed), azimuth, start and stop of each thruster).

Manual position control system: A system with centralised manual position control and automatic heading control, referred to as a joystick system.

Manual thruster control system: A system which provides an individual control lever for each thruster.

Phase back: A method utilised to temporarily reduce power consumption following an event, to stabilise the power plant and avoid a black-out.

Redundancy Design

Autonomous system: An autonomous system is a system that can control and operate itself, independently of any control system or auxiliary systems not directly connected to it. In the DP context, Autonomy is the ability of main machinery such as generators and thrusters to make themselves ready for DP without the need for hierarchical control. The use of the word autonomy here should not be confused with its use in applications such as self-driving ships and vehicles.

Bus tie breaker: A device connecting/disconnecting switchboard sections.

Closed bus: Closed bus often describes an operational configuration where all or most sections and all or most switchboards are connected together, that is, the bus-tie breakers between switchboards are closed. Also called joined bus, tied bus or closed-ring. The term closed bus/closed bus-ties refers to a configuration in which the power plant is operated as a common power system. An alternative to closed bus is open bus, sometimes called split bus or split ring.

Common cause failure: In the context of DP system, the term "common cause failure" is used to describe a failure which causes a malfunction in more than one redundant equipment group. Each redundant group may malfunction in a different way. For example, an over excitation failure in one generator causes another to trip on field failure.

Common mode failure: A failure that occurs when events are not statistically independent, when one event causes multiple systems to fail. In terms of DP system, the term "common mode failure" is a subset of common cause failure in which the redundant elements malfunction in the same way. E.g., a short circuit fault on a common power distribution system may cause a voltage dip in all redundant equipment groups which causes all power consumers to malfunction and trip on under voltage.

Critical redundancy: Equipment provided to support the worst-case failure design intent.

Cross connection or common point (X Groups): Common points, components or systems that are shared by more than one independent group or redundant group may be assigned to a common X-Group of components, these common components or common points between the redundant or independent groups are fault propagation paths with the potential to defeat the redundancy concept. Cross connections are common points introduced to achieve certain objectives, such as providing backup power, transferring power or data between redundant groups, or facilitating maintenance activities. Such cross connections are fault propagation paths with the potential to defeat redundancy. Common points and cross connections can result in fault propagation pathways leading to failures exceeding the worst-case failure design intent if not adequately mitigated by appropriate compensating provisions. Note that all cross connections are common points but not all common points are cross connections. The design of compensating provisions should take this into account.

Compensating provisions: Compensating provisions are applied to common points and cross connections to mitigate the effects of fault propagation. They may include protective functions, such as over current protection and attributes such as fault ride through capability. The effectiveness of compensating provisions should be validated by testing.

Differentiation: A method to avoid common mode failures by introducing a change in the characteristics of redundant systems based on the same principle, e.g., use of Inertial Aided Navigation (IAN) on one of the two redundant GNSS systems.

Diversity: The property of introducing differences into redundant elements to avoid common mode, common cause failures. Different levels of diversity are possible such as specifying different manufacturers for redundant GNSS systems. Even greater diversity can be achieved through orthogonality which requires redundant elements to operate on different principles.

Fail-safe: The system is to return to a safe state in the case of a failure or malfunction and this should not result in a loss of position or heading.

Hidden failure: A failure that is not immediately evident to operations or maintenance personnel, and which has the potential for failure of equipment to perform on-demand functions, such as protective functions in the power plant and switchboards, standby equipment, backup power supplies or lack of capacity or performance.

Independence: A system that can operate without the assistance of central control or other systems or subsystems. In this book, it is mainly in reference to main machinery such as generators and thrusters. Auxiliary and control functions are to be provided in a manner that makes the machinery as independent as practical to minimise the number of failures that can lead to the loss of more than one main item of machinery.

Non-critical redundancy: Equipment provided over and above that required to support the worst-case failure design intent. Its purpose is to improve the reliability and availability of systems.

Physical separation: With reference to DPS-3 vessels, fire and watertight subdivisions are required to support the worst-case failure design intent in respect of DPS-3 failure criteria.

Redundancy: Ability of a component or system to maintain or restore its function, when a single fault has occurred. Redundancy can be achieved for instance by the installation of multiple components, systems or alternative means of performing a function.

Redundancy concept: The means by which the worst-case failure design intent is achieved. It is to be documented as part of the preliminary design process.

Redundant groups (subsystems): Two or more component groups each of which is capable of individually and independently performing a specific function.

Segregation: Refers to systems or equipment intended to provide redundancy. Reduces the number of connections between systems to reduce the risk that failure effects may propagate from one redundant system to another.

Single fault: The single fault is an occurrence of the termination of the ability to perform a required function of a component or a subsystem in the DP system. For vessels with DPS-3 notation, the loss of any single compartment is also to be considered a single fault.

Single fault tolerance: The ability of a system to continue its function, following a single fault, without unacceptable interruption.

Orthogonality: With reference to redundant systems, the secondary means of providing a function should be based on completely different principles to reduce the risk of common mode failures. (e.g., Gyros-spinning mass versus Fibre Optic Gyros (FOG), anemometers (ultrasonic versus mechanical).

Performance and Operation

Activity-specific operating guidelines (ASOG): Guidelines on the operational, environmental and equipment performance limits for the location and specific activity. (For drilling operations, the ASOG may be known as the Well-Specific Operating Guidelines (WSOG)).

Availability: (IEC 191-02-05) The ability of an item to be in a state to perform a required function under given conditions at a given instant of time or over a given time interval, assuming that the required external resources are provided. This ability depends on the combined aspects of the reliability performance, the maintainability performance and the maintenance support performance.

Fault ride through: The capability of a vessel's DP system power plant and control systems (for a closed-bus operation) to recover from the effects of a fault anywhere on the main and distribution systems (e.g., on a main switchboard bus segment or on a major feeder circuit) without significant malfunction before the bus-tie breaker opens to isolate the faulted side of the system during a "limited" time or voltage drop. "Limited" time is to be defined by the designer in order to make the system stable. The severe voltage excursions associated with the operation of over current protection is a very typical example of a condition which requires ride through capability.

Loss of position and/or heading: The vessel's position and/or heading is outside the limits set for carrying out the industrial activity in progress.

DP capability/station keeping analysis: A theoretical calculation of the vessel's capability to maintain position under particular conditions of wind, waves and current from different directions. It is to be determined for different thruster combinations, (e.g., all thrusters, loss of most effective thrusters, WCF).

Maintainability: (IEC 191-02-07): The ability of an item under given conditions of use, to be retained in, or restored to, a state in which it can perform a required function, when maintenance is performed under given conditions and using stated procedures and resources within a stated time.

Operability: Service operability performance (IEC 191-19-06): The ability of a service to be successfully and easily operated by a user.

Resilience: The ability of a system to withstand a failure and to continue operations following failure. It may include the ability to recover from a failure without suffering significant damage.

Spinning reserve: The difference between the total power from immediately available sources (such as running generators, Energy Storage Systems, etc.) and the consumed power.

Station keeping: Maintenance of a desired position and/or heading or track within the normal excursions of the control system and inside the limits set for carrying out the DP activity in progress, under the defined environmental conditions (e.g., wind, waves, current, etc.).

System flexibility: The ability of a system to adapt to internal or external changes to allow its primary mission to be satisfied.

Time to safely terminate: This time is calculated as the amount of time required in an emergency to physically free the DP vessel from its operational activity following a DP abort status and allowing it to be maneuverer clear and to proceed to safety.

Documentation

Where one of the classification notations described in this chapter is requested, the following plans, data and documentations are to be submitted for review as applicable. These documentation requirements are in addition to the requirements for the vessel's mandatory classification notations. The requirements for the extent of the documents are given in the related chapters of this book. The following symbols are used in this book for the type of review of the documents:

R: Documents to be reviewed.

I: Documentation for information and verification for consistency with related review.

OB: Documentation which needs to be kept onboard.

Dynamic Positioning System

Documents listed in this section are mainly related to the global functionality, performance and characteristics of an integrated DP system.

- DP operations manual (**R**, **OB**).
- System description including a block diagram showing how the various components are functionally related (**R**).
- Details of the DP alarm system and any interconnection with the main alarm system (**R**).
- General Arrangement drawing for DP control station including control console, control panel, layout of Navigation bridge deck, list of equipment, etc. (**R**).
- Failure modes and effects analysis (FMEA, proving trials) (**DPS-2**, **DPS-3**) (**R**, **OB**).
- DP station keeping capability analysis including environmental force calculation, thruster force calculation and capability polar plots for normal operational case and for post-worst-case failure operational case. The capability plots are to represent the environmental conditions in the area of operation and the mission-specific operational condition of the vessel (**R**).
- DP system and periodic verification and testing plan (**R**, **OB**).
- Cable routing layout (**DPS-3**) (**R**).
- Details of the emergency stop system for thrusters (**R**).
- Specification of safe operating envelope of environmental conditions for DP operation (**I**).
- Basic design of DP system redundancy (**DPS-2**, **DPS-3**) (**R**).
- Description of vessel emergency shutdown (ESD) system if applicable (**R**, **OB**).
- Description of emergency disconnecting system (EDS) if applicable (**I**).
- Planned inspection and maintenance (**I**, **OB**).
- Fault ride through test plan (**R**).

Power System

Documents listed in this section are mainly related to the power system including all components and subsystems necessary to supply the DP system with power.

- Electrical power generation system (**R**, **OB**).
- Power distribution system (**R**, **OB**).
- Load analysis (**R**, **OB**).

- Auxiliary system distribution (**R**, **OB**).
- Uninterruptable power supply system (**R**, **OB**).
- Description of bus-tie breaker protective functions where applicable (**R**, **OB**).

Thruster System

- Thruster system arrangement (**R**).
- Thruster control system (**R**).
- Thrust output and power input curves (**I**).
- Thruster auxiliary system (**R**).
- Thruster monitoring system (**R**).
- Description of emergency stop system for thrusters (**R**).
- Thrust response time for thrust and direction changes (**I**).
- Thrust reductions due to interaction effects (**I**).
- Manufacturer test procedure (**R**).
- DP trial test procedure (**R**, **OB**).
- Manual for operation and maintenance (**I**, **OB**).
- Thruster Fail-Safe and Verification and Validation documents (e.g., within or part of FMEA) (**R**, **OB**).

DP Control System

- DP Control system, scope and arrangement (**R**).
- Details of the position reference system and environmental monitoring systems (**R**).
- Details of the consequence analyser (**DPS-2**, **DPS-3**) (**R**).
- Certification of suitability of control equipment for the marine atmosphere (**I**, **OB**).
- Control system functional description (**I**, **OB**).
- Manufacturers' equipment operation documents (**R**).
- Software quality plan (**I**).
- FMEA of DP control system (**DPS-2**, **DPS-3**) (**R**, **OB**).
- Manufacturer test procedure (**R**, **OB**).
- DP trial test procedure (**R**, **OB**).
- Manual for operation and maintenance (**I**, **OB**).

Certification

Control and Monitoring System Equipment

Control and monitoring (alarms and instrumentation) system equipment used in a DP control system to be assigned a DPS notation is to be certified for suitability in the marine environment. Hydraulic and pneumatic piping systems associated with the thruster system are to be subjected to pressure tests at 1.5 times the relief-device setting using the service fluid in the hydraulic system and dry air or dry inert gas for pneumatic systems as testing media. The tests are to be carried out in the presence of a Class approved Surveyor.

Components and Subsystems

The components and subsystems associated with the dynamic positioning system are to be certified or type approved according to the vessel's mandatory classification notation.

Contents

Abbreviations and Acronyms

ABS	American Bureau of Shipping
API	American Petroleum Institute
ASOG	Activity Specific Operating Guidelines
AVR	Automatic Voltage Regulator
CAM	Critical Activity Mode
CAMO	Critical Activity Mode of Operation
CPP	Controllable Pitch Propeller
DGNSS	Differential Global Navigation Satellite System
DP	Dynamic Positioning
DPCS	Dynamic Positioning Control Systems
DPO	DP Operator
ECR	Engine Control Room
EDS	Emergency Disconnect System
EGP	Enhanced Generator Protection
EHS	Enhanced System
ESD	Emergency Shut Down
ETD	Energy Transformation Device
FMEA	Failure Mode and Effect Analysis
GLONASS	Global Navigation Satellite Systems
GNSS	Global Navigation Satellite System
GPS	Global Positioning System
HSC	High Speed Craft
HVAC	Heating, Ventilating and Air Conditioning
IMCA	International Marine Contractors Association
IMO	International Maritime Organisation
ISQM	Integrated Quality Management Guide
KVAR	Kilovolt Amperes, Reactive
KW	Kilowatt
MODU	Mobile Offshore Drilling Unit
MRU	Motion Reference Unit
MTS	Marine Technology Society

MVR	Marine Vessel Rules
OB	Onboard
OCIMF	Oil Companies International Marine Forum
PMS	Power Management System
PRS	Position Reference System
RPM	Revolution Per Minute
SKP	Station Keeping Performance
SSV	Software System Verification
TAM	Task Appropriate Mode
UPS	Uninterruptable Power Supply
WCF	Worst-Case Failure
WCFDI	Worst-Case Failure Design Intent
WSOG	Well-Specific Operating Guidelines

List of Figures

List of Tables

Chapter 1
Dynamic Positioning System Design

1.1 General

Dynamic Positioning (DP) systems are advanced control systems used in ships and offshore structures to automatically maintain their position and heading without the need for traditional anchoring. DP systems are crucial for various applications, such as offshore drilling, subsea construction, and vessel operations in challenging environments. Here is an overview of the key aspects involved in designing a dynamic positioning system:

(1) Sensors:

  (a) Position reference Systems (PRS). DP systems rely on accurate position information. Common PRS include GPS, DGPS (Differential GPS), GLONASS, and other satellite-based systems. In addition to global navigation satellite systems, other sensors such as taut-wire systems, acoustics, radar, and laser-based systems can be used for precise positioning.
  (b) Motion sensors. Gyrocompasses, accelerometers, and other motion sensors provide data about the vessel's movements, including roll, pitch, and yaw.

(2) Control algorithms. DP systems use sophisticated control algorithms to process data from sensors and adjust the vessel's thrusters or propulsion systems to maintain the desired position and heading. Proportional-Integral-Derivative (PID) controllers are commonly used, but more advanced control strategies, such as model predictive control, are also employed for better performance.
(3) Thrust systems. The DP system controls the thrust and steering systems of the vessel. Thrusters can be azimuth thrusters, tunnel thrusters, or other propulsion systems. Redundancy is often built into the thrust system to ensure reliability. This may involve multiple thrusters distributed around the vessel.

F. Karkori, *Dynamic Positioning Systems*, Springer Series on Naval Architecture, Marine Engineering, Shipbuilding and Shipping 21,
https://doi.org/10.1007/978-3-031-59173-0_1

(4) Power systems. Reliable power systems are essential for DP operations. Redundancy is typically built into power generation and distribution systems to ensure continuous operation even in case of equipment failure.
(5) Human–machine interface (HMI). A user-friendly interface is crucial for operators to monitor and control the DP system. The HMI provides real-time information on the vessel's position, heading, and system status. Alarms and alerts are integrated into the interface to notify operators of any issues or deviations from the desired position.
(6) Redundancy and safety. DP systems often incorporate redundancy in critical components to enhance safety. This includes duplicate sensors, multiple control computers, and backup power systems. Safety features such as automatic position mooring, emergency shutdown procedures, and collision avoidance systems are integrated into the DP system.
(7) Testing and simulation. DP systems are rigorously tested through simulation before being deployed in real-world operations. Simulations help validate the system's performance under various conditions and ensure that operators are adequately trained.
(8) Regulatory compliance. DP systems must comply with international standards and regulations. Classification societies like the American Bureau of Shipping (ABS) and Det Norske Veritas (DNV) and Germanischer Lloyd (GL) provide guidelines for DP system design and certification.

Designing a DP system involves a multidisciplinary approach, combining expertise in control systems, marine engineering, navigation, and computer science. Additionally, ongoing maintenance and periodic system updates are crucial to ensure the continued reliability and effectiveness of the DP system.

This chapter provides an overview of the general requirements for vessels fitted with DP systems, and subsequently qualify for a **DPS**-series Class notation. More detailed requirements for each subsystem are given in Chaps. 2, 3, 4 and 5 respectively. Chapter 7 of this book provides the requirements for enhanced system notations (**EHS**) and Chap. 8 provides the requirements for station keeping performance notation (**SKP**).

1.2 DP System Technical Requirements

1.2.1 Basic Requirements

To be approved for service, any DP system must be designed to possess a certain level of station keeping capability, reliability and redundancy. The class notation for DP Systems addresses the issue of reliability based on the redundancy and fault tolerance of the DP system. The rating for DP systems is incremental. In other words, for a vessel with the notation **DPS-0**, or **DPS-1**, a loss of position and/or heading (station keeping) may occur in the event of a single fault. For a vessel with the notation **DPS-2**, a loss of position and/or heading (station keeping) will not occur in the event of a single fault in any active component or system, but for the total loss of a compartment or compartments. For a vessel with the notation **DPS-3**, a loss of position and/or heading (station keeping) will not occur in the event of a single fault in any active or static component or system, including the complete loss of a compartment due to fire or flood. The redundant components and systems are to be immediately available and with such capacity that the DP operation can continue for such a period that the work in progress can be terminated safely. The period for safely terminating a work in progress must be specified by the vessel owner. The selection of the level of DP Class notation of the vessel for a particular operation is outside the scope of this book. Some Coastal States impose minimum DP Equipment Class requirements for activities carried out within their jurisdiction. The station keeping capability is normally defined and presented by the limiting environmental conditions under which the vessel can maintain the position and heading, both in intact and post worst-case failure conditions. The limiting environmental conditions and operational modes for a DP vessel are to be defined by the vessel owner.

Note: *some of the requirements outlined in this book were sourced from the MTS Document titled "MTS, Technical and Operational Guidance (TECHOP) TECHOP (D-01-Rev1-Jan21)", which addresses C3EI2 to eliminate Single Point Failures (C3EI2 - Cross-connections, Commonality, External Interfaces and Influences) January 2021. The text for those requirements is given in italics.*

1.2.2 DPS-Series Notations

For the notation **DPS-0** and **DPS-1**, there is no requirement for redundant systems. For the **DPS-0** notation, the DP system is to have a manual position control system. For the **DPS-1** notation, the DP system is to include an automatic DP system and a manual position control system. A vessel with **DPS-2** or **DPS-3** notation is required to have an automatic DP system, manual position control system and to be single fault tolerant. The single fault tolerance is to be achieved by the design of redundant systems. The station keeping capability after a single fault is to be achieved by providing control, electric power and thrust, without reliance on operator intervention. For **DPS-1 +**, **DPS-2 +** or **DPS-3 +** notation, transverse thrust force and yaw moment are not to be solely dependent on the propellers and rudders. The thruster configuration is to be arranged so that they can provide transverse force and yaw moment for station keeping. For the **DPS-2** notation, common static components may be accepted in systems which do not immediately affect position keeping capabilities upon failure (e.g. ventilation, and seawater systems not directly cooling running machinery). Normally, such static components will not be considered to fail where adequate protection from damage is demonstrated. For the **DPS-2** notation, a single fault may include:

- Any active component or system (generators, thrusters, switchboards, communication networks, DP control computers, sensors, remote controlled valves, etc.).
- Any normally static component (cables, pipes, manual valves, etc.) that may immediately affect position keeping capabilities upon failure or is not properly documented with respect to protection.

For the **DPS-2 +** notation, a single fault may include:

- Any of the items listed above for **DPS-2**.
- Coolers and filters.
- Fuel oil service tanks and supply pipes to engines.
- Short-circuit of bus-bars for switchboards, where applicable.
- Fault tolerance to short circuits and ground faults is to be validated by testing, where applicable.

For the **DPS-3** notation, a single fault may include:

- Any of the items listed above for **DPS-2**, and any normally static component is assumed to fail.
- Any components in any one watertight compartment from flooding.
- Any components in any one fire subdivision from fire.

1.2.3 Consideration of Redundancy

For the DP system to have appropriate redundancy, the redundancy is to have two or more items of equipment or system required to perform a function so that the redundant unit can take over from the failed unit without unacceptable interruption of the function. Redundant components and systems are to be immediately available for use without needing manual intervention from the operators and with such capacity that the DP operation can be continued for such a period that the work in progress can be terminated safely. The transfer of control is to be smooth and within acceptable limitations of the DP operation(s) for which the vessel is designed. In general, a full stop and restart of the system does not comply. Automatic start of equipment may be accepted as contributing to redundancy only if they can be tested to prove that they can be brought into operation automatically before position and heading keeping performance is degraded. The redundancy depends on systems being available in both number and capacity to produce the required DP capability after worst-case failure. Independence of redundancy groups is to consider all technical functions. The redundancy design can consist of two fully redundant power and thruster systems each capable of maintaining position and heading if the other fails. The design can also make use of multiple systems each providing partial redundancy such that the vessel can maintain position with all combinations of independent systems that survive any defined fault. The redundancy design is to provide suitable combinations of available systems following any defined fault. The transfer of failures between redundant subsystems is to be prevented by effective compensating provisions. Efficacy of compensation provisions is to be demonstrated by validation testing. The failure of redundant components or systems is to be revealed by alarms. The possibility of hidden failures is to be minimised through periodic testing.

1.3 Basic Design of Redundancy and Redundancy Concept

Basic design of redundancy and the redundancy concept of the DP system are to be developed at the early design stage and are to be submitted to Class for review. The documentation is to include at least the following:

- General arrangement of DP system.
- Worst-case failure design intent (WCFDI).
- Limiting environmental conditions for DP operation.
- Power plant configurations for DP operation.
- Allowable loss of power sources after one failure.
- Allowable number of failed thrusters.
- Time period for safely terminating a DP operation after each considered single fault.

The documentation is to be reviewed during the design process is to confirm if the worst-case failure design intent is met. The worst-case failure design intent is to describe the minimum amount of propulsion and control equipment remaining operational following the worst-case failure and is to be used as the basis of the design development.

1.3.1 Redundancy Design Analysis

Redundancy is to be verified through Failure Mode and Effects Analysis (FMEA) considering relevant failure modes.

1.3.2 Cross Connections or Common Points

For **DPS-2** and **DPS-3** notations, connections between otherwise redundant and separated systems or redundancy groups are to be kept to a minimum and made to fail to the safest condition. Failure in one system is not to propagate to another redundant system. The DP FMEA is to identify, assess, and address cross connections. Cross connections are to be addressed by fail-safe designs and conditions, protecting functions, and testing. Tests for identified cross connections are to be included in the

DP FMEA Trials. The FMEA is to identify and analyse for hidden failures which will result in a loss of redundancy. Compensating provisions may include protective functions, isolation strategies, alarms, monitoring, and periodic testing or procedures. The term ‘cross connections’ means physical connections between redundant equipment groups. Cross connections can also be created by the colocation of equipment supplied from redundant power distribution systems. In **DPS-3** designs, these connections are made by considering the effects of fire and flooding in a common compartment.

1.3.3 Hidden Failure Monitoring

Hidden failure monitoring is to be provided on all devices where the FMEA shows that a hidden failure will result in a loss of redundancy.

1.3.4 Fault Propagation Paths

Cross connections are potential fault propagation paths that may spread failure effects in one redundant equipment group to another. Such coupling or spreading condition may result in the failure of both systems accompanied by failure effects of a severity greater than that of the Worst-Case Failure Design Intent (WCFDI).

1.3.5 Marine Auxiliary Services

Cross connections in marine auxiliary systems are allowed in some DP notations and there may be no separation of pipework in some designs. Such designs can be vulnerable to common mode failures associated with contamination, leakage or aeration of the fluids they transport. Such vulnerabilities are usually managed with effective enforcement of procedural barriers. Full separation of marine auxiliary systems is desirable. Cross connections for maintenance purposes can be accepted with mitigation measures in place.

1.4 Physical Separation

For a vessel with a **DPS-3** notation, the redundancy groups are to be separated by bulkheads and decks that are fire-insulated by A-60 class divisions. The separation is to be watertight if below the damage water line. Cables for redundant equipment or systems are not to be routed together through the same compartments. If not avoidable, cables for redundant systems are to be separated by at least one A-60 barrier. Thus, cables associated with one system could be run in an A-60 class duct with the other system running through the compartment or both sets of cables routed through separate ducts. Redundant piping system (i.e., piping for fuel, cooling water, lubrication oil, hydraulic oil, etc.) is not to be routed together through the same compartments. If not avoidable, such pipes are to be separated by at least one A-60 barrier. Thus pipe-work associated with one system could be run in an A-60 class duct with the other system running through the compartment or both sets of pipe-work routed through separate ducts.

1.5 DP System Equipment Requirements

To meet the requirements for a **DPS**-series notation, the minimum number of subsystems and components and the redundancy for: power system, thruster system and DP control system are provided in Table 1.1. The detailed requirements for each subsystem and the arrangement of the equipment are given in Chaps. 2, 3, 4 and 5, respectively.

1.6 Station Keeping Performance

A DP system is to be designed to be able to maintain the position and heading under the specified limiting environmental conditions. The design environment is to be specified by the Vessel Owner. A vessel with a **DPS-0** or **DPS-1** notation is to be able to maintain position and heading under specified environmental conditions with all thrusters intact. A vessel with a **DPS-2** or **DPS-3** notation is to be able to maintain position and heading under specified environmental conditions during and after any single fault of the DP system. The station keeping performance assessment is to be carried out for:

- Design environmental conditions.
- Normal operational condition.
- Standby condition (where applicable).
- All thruster intact condition.
- Onset of Worst-Case Failure as defined in the Redundancy Concept Worst-Case Failure Design Intent.
- Other conditions, such as the failure of the most effective thruster or least effective thruster, may also be included in the assessment.

The environmental load calculations, including those for wind, current and wave drift forces, are to follow suitable procedures. API RP 2SK and the Class specific rules for mobile offshore units may be used for environmental load calculation. Thruster degradation due to interactions is to be considered in the station keeping capability

Table 1.1 Summary of DP system requirements for DPS notations (4)

Subsystem or component	Equipment		Minimum requirements for each Classification notation			
			DPS-0	DPS-1(7)	DPS-2(8)	DPS-3(5)(9)
Power system	Generators and prime movers		Non-redundant	Non-redundant	Redundant	Redundant, in separate compartments
	Main switchboard		1	1	1 with bus-tie	2 with bus-ties, in separate compartments(5)
	Bus-tie breaker		0	0	1	2
	Distribution system		Non-redundant	Non-redundant	Redundant	Redundant, in separate compartments
	Power management(2)		No	No	Yes	Yes
Thrusters	Arrangement of thrusters		Non-redundant	Non-redundant	Redundant	Redundant, in separate compartments
Control system	DP control: number of control computers		0	1	2	2 + 1 in backup control station
	Manual position control: joystick with auto heading		Yes	Yes	Yes	Yes
	Manual thruster control		Yes	Yes	Yes	Yes
	Position reference systems		1	2	3	2 + 1 in backup control station
	Sensors:	Wind	1	2	3	2 + 1 in backup control station
		MRU(3)	0	1	3	2 + 1 in backup control station
		Vessel heading(6)	1	2	3	2 + 1 in backup control station
	UPS		0	1	2	2 + 1 in separate compartment
Backup control station for backup unit			N/A	N/A	N/A	Yes

(continued)

Table 1.1 (continued)

Subsystem or component	Equipment	Minimum requirements for each Classification notation			
		DPS-0	DPS-1[7]	DPS-2[8]	DPS-3[5][9]
Consequence analyser		No	No	Yes	Yes
FMEA		No	No	Yes	Yes

Notes: (1) **DPS-0** is an ABS system class. It applies to a manual position control system fitted with automatic heading control and with a free-standing position reference system. **DPS-1**, **DPS-2** and **DPS-3** are in line with IMO equipment class 1, class 2 and class 3, respectively. (2) If all thrusters are direct diesel drive, a power management system is not required. (3) If position reference systems are dependent on correction of the measurements for roll and pitch noise, their associated MRUs are required. (4) For enhanced system (**EHS-E**, **EHS-P**, **EHS-F** and **EHS-C**), additional information is provided in Chap. 8, Table 8.1. (5) Where "separate compartments" is indicated, the equipment is to be located in separate compartments arranged to support the worst-case failure design intent in respect of **DPS-3** failure criteria. (6) Where three (3) vessel heading sensors are required, one of these may be a THD (Transmitting Heading Device), such as a GPS-Satellite-Compass. This is to have type approval for compliance with the Performance standards for Marine Transmitting Heading Devices (THDs), as referenced by SOLAS Chapter V, by a recognised certification agency/ organisation. (7) **DPS-1 +** notation may be assigned for DP system meeting **DPS-1** requirements plus 1 stern thruster. (8) **DPS-2 +** notation may be assigned for DP system meeting **DPS-2** requirements plus 2 stern thrusters and redundancy of static components. (9) **DPS-3 +** notation may be assigned for DP system meeting **DPS-3** requirements plus 2 stern thrusters

assessment. Other effects due to industrial activities, such as pipe tension, heavy lift, hawser tension, riser load, etc., are to be considered in the station keeping capability analysis. The effect of the vessel's industrial activities on available power for the DP system is also to be considered in the assessment. Station keeping capability can be presented using traditional DP capability plots or using thrust utilisation plots. The station keeping capability analysis report is to be submitted for review. The following information is to be included in the report:

- Vessel general arrangement and main particulars.
- Projected areas for wind and current.
- Available power for DP system.
- Thruster arrangement, maximum thrust.
- Thruster degradation.
- Wind, current and wave force coefficients.
- Wind and wave relationship used where applicable.
- Description of analysis method.
- DP capability plots.
- Thrust utilisation under the specified limiting environmental conditions where applicable.

1.7 Essential Non-DP Systems

Essential non-DP systems are essential systems which are not directly part of the DP systems, such as common fire suppression systems, engine ventilation systems, heating, ventilation and air conditioning (HVAC) systems, and emergency shutdown systems, that in the event of failure could cause a failure in the DP system. The redundancy concept for the DP system is to be followed through to these systems so that actions or failures initiated by these systems do not cause consequences that exceed the worst-case failure design intent. The actions initiated by these systems are to be scaled to the detected threat level and are to be addressed in the FMEA of the DP system. The design and operation of ESD and Fire and Gas Safety shutdown systems are to meet the following:

- be in line with the DP redundancy concept of the vessel.
- subjected to a system engineering approach.
- subjected to documented proving trials.

Nuisance alarms, fire and gas safety shut down alarms and non-explosive gas influx related triggers, are not to result in an automatic shutdown or loss of thrusters. Default operation of emergency shutdown (ESD) and fire and gas safety shutdown systems are to be in advisory mode (initiating alarms). Operating in supervisory mode with the potential of shutting down DP critical equipment due to non-explosive gas influx related triggers or spurious triggers is to be avoided. Any decision to operate in supervisory mode is to be based on verification of the systems engineering approach and proving trials to demonstrate that the system is resilient to shutdowns triggered due to spurious events (e.g., sensor failures, input output card failures (benign and aggressive failures)).

1.7.1 Emergency Shutdown System and DP Redundancy

The DP system is not to be affected by failures of the emergency shutdown systems. For **DPS-2** that will include any failure in the shutdown systems and for **DPS-3** it will include the effect of fire on associated cabling and piping in any compartment. The effect on the DP system due to any reasonable act of maloperation is to be considered. This means that the shutdown systems are treated in the same way as any other part of the vessel's DP system that meets the same redundancy requirements and have their failure modes analysed in the FMEA of the DP system. In spaces such as thruster rooms and auxiliary machinery spaces, an immediate effect on DP systems due to a spurious shutdown of ventilation is to be prevented and to be confirmed during FMEA proving trials.

1.8 Fire Protection

Fixed firefighting systems may include CO^2 or other fire suppressant agents such as water mist. These systems are to be arranged in a manner that supports the overall divisions in the DP redundancy concept. Where possible, the release of the fire suppressant is to only affect one redundant machinery group.

1.9 Fuel Quick Closing Valves

These valves are provided to allow rapid isolation of fuel supplies in emergency situations. The valves are to be provided in line with the requirement of the vessel's mandatory classification notations. They are to be arranged in a manner that allows fuel to be isolated to only one redundant machinery group without affecting the operation of any others. Due consideration is to be given to verify that the arrangement does not introduce a common point which could defeat the DP redundancy concept. Efficacy of compensating provisions/mitigating measures to address any vulnerability due to the common point is to be validated by testing.

1.10 Alarms, Monitoring and Instrumentation

The main purpose of the alarm and monitoring system is to give the DP operators the abnormal condition and status information they require to maintain safe and efficient operation of the system. Information relating to power management, propulsion, ballast control, HVAC, etc., are to be available. The displays and alarms as specified in Table 1.2 are to be provided at DP control stations, as applicable. Table 1.2 provides the summary of minimum basic instrumentation at the main DP control station. Additional instrumentation may be necessary. The detailed information for displays and alarms for the DP system's components and subsystems are given in Chaps. 2, 3, 4, and 5 respectively.

1.11 Communications and DP Alert System

1.11.1 Communications

Communication is a key management tool during execution. This is to be incorporated in the design phase. At least one means of voice communication is to be provided between each DP control station and the navigation bridge, the engine control position, any other relevant industrial function control centres associated with DP and

Table 1.2 Summary of minimum instrumentation at DP control station

System	Monitored parameters	Alarm	Display
Thruster power system	Engine lubricating oil pressure–low	X	
	Engine coolant temperature–high	X	
	CPP hydraulic oil pressure–low and high	X	
	CPP hydraulic oil temperature–high	X	
	CPP pitch		X
	Thruster RPM		X
	Thrust direction		X
	Thruster motor/semiconductor converter coolant leakage	X	
	Thruster motor semiconductor converter temperature		X
	Thruster motor short circuit	X	
	Thruster motor exciter power available		X
	Thruster motor supply power available		X
	Thruster motor overload	X	
	Thruster motor high temperature	X	
	Thruster operation (on-line/off-line)		X
Power distribution system	Status of automatically controlled circuit breakers		X
	Bus bar current and power levels		X
	High power consumers – current levels		X
	Status of power management system	X	X
	Spinning reserve		X
System performance	Excursion outside operating envelope	X	
	Control system fault	X	
	Position sensor fault	X	
	Vessels target and present position and heading		X
	Wind speed and direction		X
	Selected reference system		X
Specific requirements for **DPS-2 & DPS-3**	Thruster location (pictorial)		X
	Available thrust used and thrust vector		X
	Available thrusters on stand-by		X
	Consequence analyser alert	X	
	Position information of individual position reference systems connected		X

any location required by the vessel's mandatory classification notation. The voice communication system is to be powered by a battery or an uninterruptible power supply system sufficient to operate the system for at least 30 min. The communication systems are to be located within easy reach of the DP operator at the DP control stations.

1.12 DP Alert System

DP alert statuses in operation must be clearly defined and are, at a minimum, to have the following three levels.

(1) Normal operational status.
(2) Degraded DP status.
(3) DP emergency status.

A system of visual and audible alarms is to be provided at each DP control station, on the navigation bridge and at the propulsion engine control position or centralised control station, if fitted. The alarms are to be capable of being manually activated from the DP control stations (including DP back-up control station, if fitted) to indicate DP operational status. Where such an alert system is not easily included the means of clear communication of the statuses are to be agreed before commencement of operations. Chapter 9 of this book provides more detailed information for specific vessel types. The guidance provided by IMCA M 103 "*Guidelines for Design and Operation of Dynamically Positioned Vessels*" and by MTS "*DP Vessel Design Philosophy Guidelines*" can be used for the design of DP alert system.

1.13 Failure Mode and Effects Analysis

FMEA is required for **DPS-2** and **DPS-3** notations and any notations that require **DPS-2** or **DPS-3**. In general, two FMEAs are to be considered, one covering the main DP control systems and the other for all other systems onboard related to DP operations. The purpose of the FMEA is to indicate whether or not the DP system meets the requirements of the relevant DP notation and complies with the vessel's WCFDI. The DP FMEA may be performed based on the guidance provided throughout this book, in addition to IMCA M 166, IEC 60,812, "*DP Vessel Design Philosophy Guidelines*" by MTS, Annex 4 of IMO High Speed Craft Code or equivalents. The objective of the DP FMEA is to, at a minimum, include the following:

- Identify and provide recommendations to eliminate or mitigate the effects of all single faults and common mode failures in the vessel DP equipment which, if any occurs, would cause total or partial loss of station keeping capability.
- Demonstrate effective redundancy.

- Identify potential "hidden" failures with the potential to defeat the DP redundancy concept.

Results of the DP FMEA are to be verified during FMEA proving trials. The DP FMEA test program is to be developed based on the analysis results, analysis method and assumptions. The relevant test program is to be provided to Class for review and approval. DP FMEA and DP FMEA test programs are to be kept onboard and they are to be updated to cover subsequent alterations to the DP system hardware or software.

1.13.1 Failure Mode Analysis

For a **DPS-2** or a **DPS-3** notation, loss of position and/or heading (station keeping) is not allowed to occur in the event of a single fault. Single fault includes, but is not limited to the following:

- All redundant components, systems or subsystems.
- A single inadvertent act of operation (ventilation, fire suppression, ESD) where applicable and if such an act is reasonably probable.
- Hidden failures (such as protective functions on which redundancy depends) where applicable.
- Common failure modes.
- Cross connections or common points.
- Governor and AVR failure modes where applicable.
- Main switchboard control power failure modes.
- Bus tie protection where applicable.
- Power management system.
- DP control system input and output arrangement.
- Position reference processing.
- Networks.
- Communication failure.
- Automatic interventions caused by external events, when found relevant (e.g., automatic action upon detection of gas).

The failure mode analysis is also to include:

- The most predictable cause associated with each failure mode.
- The method of detecting that the failure has occurred.
- The effect of the failure upon the rest of the system's ability to maintain position.
- An analysis of possible common failure modes.
- Benign and aggressive failure modes.

Where parts of the system are identified as non-redundant and where redundancy is not possible, these parts are to be further studied with consideration given to their reliability and mechanical protection. The results of this further study are to be submitted for review. When there are more configurations for the diesel electric plant design to cope with equipment unavailability (e.g., failures or equipment taken down for maintenance), it is important that all configurations that are possible to be included in DP operations are analysed in the vessel's DP system FMEA to prove that the DP system remains redundant. Fault tolerance of the configurations is to be made visible and understood by the crew. Compensating provisions upon which the redundancy concept depends upon are to be clearly identified. Analysis is to cover alternate configurations the vessel intends to operate in (due to long term unavailability of DP related Station Keeping Equipment or equipment rendered unavailable due to maintenance requirements). An FMEA worksheet is to be compiled for each equipment failure assessment. Some pertinent aspects to be included in the worksheets are:

- System name (including main system, system, and subsystem).
- Reference drawings.
- Equipment name or number.
- Function description.
- Operational mode.
- Failure modes.
- Failure causes.
- Failure effects (including local effect and end effect).
- Failure detection.
- Corrective action.
- Severity of failure effect (providing definitions of categories of severity).

FMEA worksheet examples given in IEC 60,812, OCIMF DP FMEA Assurance Information Paper, Sect. 3.4 of IMCA M 166 or Appendix 2 of Annex 4 of *IMO International Code of Safety for High-Speed Craft* may be used.

1.14 FMEA Report

1.14.1 FMEA Analysis Report

The DP FMEA analysis report is to be sufficiently detailed to cover all the systems associated with the dynamic positioning of the vessel. The DP FMEA analysis report is to be a self-contained document including, but not limited to the following:

- A brief description of the vessel, vessel's worst-case failure design intent and whether the analysis has confirmed or disproved it.
- Definitions of the terms, symbols and abbreviations.
- Analysis method and assumptions.
- A description of all the systems associated with the dynamic positioning of the vessel and a functional block diagram showing their interaction with each other. Such systems would include the DP electrical or computer control systems, electrical power distribution system, power generation, fuel systems, lubricating

oil systems, cooling systems, backup control systems, thruster/propulsion/rudder systems (as applicable), all DP pertinent networks, etc.
- System diagrams are to be included where appropriate.
- A description of each physically and functionally independent item and the associated failure modes.
- Identification of redundancy groups for each system. This can be provided in tables and/or color-coded system diagrams and/or other clearly understood methods.
- Fail safe condition for each redundancy group as applicable.
- For each system, identification of cross-connections and common points. Each cross- connection or common point is to be analysed for propagation with description of the impact on the redundancy concept. Provisions to compensate or mitigate the effects of the cross-connection are to be documented, analysed, and validated by testing.
- A description of the effects of each failure mode alone on other items within the system and on the overall dynamic positioning system.
- Identification of all operating modes with a description of different configurations and settings associated with the operating modes. Failure modes and tests are to incorporate all operating modes.
- Analysis findings and recommendations.
- Conclusions including worst-case failure and recommended changes.
- Recommended FMEA tests.

The FMEA analysis report is to be updated after major modifications and is to be kept onboard the vessel.

1.14.2 FMEA Proving Trial Report

A FMEA proving trial procedure is to be developed as part of the FMEA study. The objective of the FMEA proving trial is to confirm the FMEA analysis findings and also to confirm that essential functions and features upon which the fault tolerance of the DP system depends are functional in so far as it is practical to do so (attributes of protection, performance, detection, power management, etc.). The proving trial report is to establish the FMEA test list and the corresponding test procedures including but not limited to the following:

- Purpose of test or failure mode (objective of the test and the attributes that are intended to be validated by the test, e.g., fault ride through capability test, etc.).
- Vessel and equipment setup.
- Test method.
- Expected results.
- Observed results.
- Failure detection.
- Failure effects.
- Outstanding or resolved action items.

- Comments.
- Witness name, signature and date for each test.

After completion of DP proving trials, the final version of DP FMEA analysis and DP proving trial report, including final analysis/conclusions based on actual results from DP testing, are to be submitted.

1.15 DP Operations Manual

For each vessel, a vessel specific DP system Operations Manual is to be prepared and submitted solely for verification that the information in the manual, relative to the dynamic positioning system, is consistent with the design and information considered in the review of the system. One copy of the Operations Manual is to be kept onboard. It is to be readily accessible at the DP control location and used by the DP operators as a reference for conducting DP operations. The DP system Operations Manual is intended to provide guidance for the DP operator about the specific DP installations and arrangements of the specific vessel. The DP Operations Manual is to include but is not limited to the following information:

- Organisation and responsibility during DP operations.
- DP system operator's and ECR operator's responsibility and manning.
- Definitions of the terms, symbols and abbreviations.
- The associated functional roles/operations for which the DPS notation is granted (e.g., Articulated Tug Barge Combination, Anchor Handling, Fire Fighting, Offshore Support, Diving, Underwater Inspection, etc.) and a description of the functions and equipment involved with maximum available electrical power to perform the specified vessel combined functions.
- A description of all the systems associated with the dynamic positioning of the vessel, including backup systems and communication systems.
- The block diagram showing how the components are functionally related.
- The restriction of maximum environmental condition for DP operation (e.g., wind speed, wave, current, tide, etc.).
- A description of the different operational modes and transition between modes.
- A functional description of each system, including backup systems and communication systems.
- Operating instructions for the normal operational mode (and the operational modes after a failure) of the DP electrical or computer control systems, manual position control system, manual thruster control system, DP equipment (thrusters, electric motors, electric drives or converters, electric generators, etc.).
- Description of fault symptoms with explanation and recommended corrective actions. This could be the reference to specific documents, such as *Well Specific Operational Guidelines* (WSOG) or *Activity Specific Operational Guidelines* (ASOG), and the locations of the documents.

- Instructions for tracing faults back to functional blocks or systems. This could be the reference to specific documents, such as functional descriptions of the systems, DP FMEA documents, and the locations of the documents.
- Description of cross-connections and common points identified in the DP FMEA along with appropriate corrective actions and compensating or mitigating provisions.
- Location check list (e.g., field arrival).
- Watch keeping check lists (during DP operation).
- Field arrival trials procedures.
- Training and drill (DP related).
- References to where more specific information can be found onboard the vessel, such as the detailed specific operation instructions provided by the manufacturer of the DP electrical or computer control systems, manufacturer's troubleshooting procedures for vendor-supplied equipment, DP FMEA and DP FMEA Trials, etc.
- DP capability plot.
- Incident report instruction and format.
- Annual tests and procedures.
- Examples of tests and procedures after modifications and non-conformities.
- Blackout recovery procedure.
- List of critical components and dependencies critical to the defence of the DP redundancy concept.

Note: *These Elements Are to Be Addressed in the Preventive Maintenance System and Vessel Specific Familiarisation for on Board Personnel.*

Chapter 2
Power System Design

The power systems which support and are employed by the DP system must be in compliance with the relevant Class Rules for that vessel's mandatory Classification notations. This chapter provides additional requirements for **DPS-2** and **DPS-3** notations in regard to redundancy and with respect to maximum single failure, as specified for each Class notation. Accordingly, IMO MSC.1/Circ.1580 states:

> For equipment class 2, the power system is to be divisible into two or more systems so that in the event of failure of one system at least one other system will remain in operation and provide sufficient power for station keeping. The power system(s) may be run as one system during operation but is to be arranged by bus tie breakers to separate the systems automatically upon failures which could be transferred from one system to another, including, but not limited to, overloading and short-circuits."
>
> For equipment class 3, the power system is to be divisible into two or more systems so that in the event of failure of one system, at least one other system will remain in operation and provide sufficient power for station keeping. The divided power system is to be located in different spaces separated by A-60 class division. Where the power systems are located below the operational waterline, the separations are also to be watertight. Bus tie breakers are to be open during equipment class 3 operations unless equivalent integrity of power operation can be accepted.

The above criteria from IMO MSC.1/Circ. 1580 are to be followed in the design of the power system for **DPS-2** and **DPS-3** systems. Alternative energy storage systems (e.g. Li-Ion batteries, supercapacitors, and flywheels) may be used as energy storage sources of power to thrusters as long as all relevant redundancy, independency and separation requirements for the relevant notation are complied with. For equipment classes 2 and 3, the available energy from such sources may be included in the consequence analysis function outlined in the section *Vessels with* **DPS-2** *notation* below when reliable energy measurements can be provided for the calculations.

F. Karkori, *Dynamic Positioning Systems*, Springer Series on Naval Architecture, Marine Engineering, Shipbuilding and Shipping 21,
https://doi.org/10.1007/978-3-031-59173-0_2

2.1 Power Generation System

2.1.1 Vessels with DPS-1 Notation

Generators and their distribution systems are, as a minimum, to have the capacity to supply sufficient power to the thrusters to maintain vessel's position within the specified operating envelope in addition to supplying industrial activities and essential ship service loads. When power is shared, power supply to industrial activities and essential ship service loads is not to affect DP operations.

2.1.2 Vessels with DPS-2 Notation

In addition to the criteria above for **DPS-1**, generators and their distribution systems are to be sized and arranged for Worst Case Failure of any bus section. Sufficient power is to remain available to supply essential ship service loads, critical operational loads and maintain the vessel's position and heading within the specified post failure operating envelope. The post failure remaining power plant is to be able to start any non-running load without the associated voltage dip causing any running motor to stall or its control equipment to drop out. Essential services for generators and their prime movers, such as cooling water and fuel oil systems, are to be arranged such that, with any single fault, sufficient power remains available to supply the essential loads and to maintain position and heading within the specified post failure operating envelope.

2.1.3 Vessels with DPS-3 Notation

In addition to the criteria above for **DPS-2**, generators and their distribution systems are to be sized and arranged in at least two compartments with A-60 and watertight boundaries so that, if any compartment is lost due to fire or flood, sufficient power is available to maintain position and heading within the specified post failure operating envelope, and to start any non-running load without the associated voltage dip causing any running motor to stall or control equipment to drop out. Essential services for generators and their prime movers, such as cooling water and fuel oil systems, are to be arranged so that, with any single fault in the systems or the loss of any single compartment, sufficient power remains available to supply the essential loads, the critical operational loads and to maintain position and heading within the specified post failure operating envelope.

2.2 Power Distribution

The switchboard is to be arranged for manual and automatic remote controls and be provided with all necessary alarms, controls and indications to allow local manual control of the power plant. The distribution system at the main power generation level is to be arranged to reflect the split in the redundancy concept. The split in the auxiliary power system is to follow the split in the main power generation system to match the worst-case failure design intent. Every UPS and battery system is to have a main power supply from an auxiliary system switchboard appropriate to the split in the redundancy concept and if a backup supply from the emergency switchboard is provided, loss of the emergency switchboard is not to prevent starting of main generators after a blackout. The status of automatically-controlled circuit breakers is to be monitored as described in Chap. 1, Table 1.2. The bus-bar current and power levels are to be monitored as described in Chap. 1, Table 1.2. An alarm is to be initiated upon failure of any of the required power supplies. Spinning reserve, the difference between online generator capacity and consumed power is to be displayed in the DP control station. The display is to be continuously available and is to be provided for individual bus sections for split-bus power arrangements.

2.2.1 Vessels with DPS-2 or DPS-3 Notations

For **DPS-2** or **DPS-3**, the switchboard is to be designed such that no single fault will result in a total black-out, including failure of all equipment in any fire and/or watertight subdivision for **DPS-3**. For **DPS-2**, a main bus bar system consisting of at least two sections, with at least one bus tie breaker between any two bus sections, is to be arranged. For **DPS-3**, each switchboard room is to be separated by watertight A-60 partitions. A bus tie breaker on each side of the partition is to be arranged. Bus ties are to be designed to prevent a fault from propagating from one bus section to another. When the DP system is designed including the configuration of closed bus tie breaker, this breaker is to be:

- A circuit breaker capable of breaking the maximum short circuit current in the connected system.
- Coordinated in relation to generator breakers to avoid total loss of main power (black-out).

Details of the bus tie breakers and generator breakers, such as protection settings, coordination study between bus tie breakers and generator breakers and other related information are to be submitted for review. Consideration is to be given to effective intelligent detecting and executing methods featuring ultra-fast acting actions by the devices, including rapid communication to other protective systems under the coordination scheme, to prevent and/or mitigate the detected fault being migrating to other parts of the switchboard. Bus bar control and protection systems are to be designed to work with both open and closed bus tie breakers. Analysis of relevant failure modes with closed bus tie configurations to be addressed in the FMEA so that the redundancy of the DP system meets the design intent. For the relevant failure modes, reference is made to the following:

- MTS "*DP Vessel Design Philosophy Guidelines*".
- IMCA M 103 "*Guidelines for the Design and Operation of Dynamically Positioned Vessels*".
- IMCA M 247 "*Guidance to Identify DP System Components and their Failure Modes*".

For **DPS-3**, in addition to the above requirements, the closed bus design is to include the following:

- Power system protection as outlined in Chapter 7, *Enhanced systems*, of this book.

Fault ride through capability. All equipment essential for dynamically positioning system are to have fault ride through capability, allowing for a short circuit condition to clear before under voltage protection is actuated. Low voltage transients during a short circuit condition are not to cause the motor starter to drop out, or other drives to fail. Refer also to the section on *Fault ride through (FRT) capability.*

2.3 Power Management System

The power management system is to be capable of operating with both open and closed bus tie breakers where applicable. For a **DPS-2** or a **DPS-3** notation, where DP operations are configured with diesel electric driven thrusters, power management systems are to be provided. Power management systems may be of an individual designed type or integrated with other switchboard/generator control systems. The PMS is to have redundancy according to the equipment class and a blackout prevention function.

- Power management system is to be capable of providing sufficient power for essential operations, and to prevent loads from starting while there is insufficient generator capacity. Facilities to disable the automatic functions: stop running generators, or to disconnect generator breakers, bus tie breakers, or thruster breakers are to be provided to allow manual operation by the DP operator when necessary.
- Consideration is to be given to techniques such as power limiting of heavy consumers, shedding of non-essential loads and temporary thrust reduction to maintain the availability of power. Total failure of the power management system is not to produce failure effects exceeding the worst-case failure design intent and to be demonstrated through FMEA.
- Power management system is to be supplied with an uninterruptible power supply system (UPS).

- A failure in the power management system is to initiate an alarm in the DP control station. When the power management is disconnected, manual operation of the switchboard.
- Loss of an online generator is not to result in the sustained overloading of the generators remaining online. If sufficient power is not available, the power management system in conjunction with "controls" of consumers is to reduce system load in a coordinated fashion to restore power balance. The restoration of power balance may be accomplished by load reduction of specific consumers, load shedding and sectionalisation of the electrical network.
- The power management control system is not allowed to remove the last online generator. The remove of the last online generator is to be determined by the DP operator.
- When the DP system is designed with a closed bus tie configuration for **DPS-2** or **DPS-3**, the power management system is to have protective measures implemented in order to provide the required integrity between the redundancy groups. The power management system is also to be able to communicate with other alternate protection systems if applicable. Analysis of relevant failure modes are to be addressed in the FMEA.
- For **DPS-3** notation, the power management system is to be arranged such that no single fault, including fire or flood in one compartment, will render all the power management systems inoperable.

If all thrusters are direct diesel engine driven, a power management system is not required. Where the electrical distribution systems are provided by a mixture of shaft generators from the main propulsion shafts and axillary generators, a power management system may be exempt provided other means of limiting and securing the power to thrusters for DP operations are in place and the main switchboard is arranged with split bus bar configurations.

2.4 Uninterruptible Power Supply Systems (UPS)

For the **DPS-1**, **DPS-2** or **DPS-3** notation, uninterruptible power supply system (UPS) is to be provided (i.e. minimum one UPS for equipment class 1, two UPSs for equipment class 2 and three UPSs for equipment class 3) for automatic DP control system and its associated monitoring and reference system. The arrangement and number of UPSs are to be in accordance with Table 2.1. The reference systems and sensors are to be distributed on the UPSs in the same manner as the control systems they serve, so that any power failure will not cause loss of position keeping ability. UPS arrangements are to support the WCFDI and the redundancy design. UPSs that do not follow the redundancy design on a vessel may result in undesired cross connections between redundancy groups and may lead to a failure which exceeds the WCFDI. Where a group of UPSs shares a common power source from a switchboard, loss of that switchboard is not to exceed the worst-case failure design intent when all

Table 2.1 Requirements for ESS

Applicable system		Reference
ESS	Lithium batteries	Class Rules for the use of lithium batteries in the marine and offshore industries
	Super capacitors	Class Rules for the use of supercapacitors in the marine and offshore industries

UPS batteries in that group are depleted. Each uninterruptible power supply system is to be capable of supplying power for a minimum of 30 min after failure of the main power supply. A fault in any UPS is to initiate an alarm in the DP control station. For the **DPS-2** or **DPS-3** notation, the power supplies of the UPSs for the main DP control system are to be provided from different sides redundancy groups. For the **DPS-3** notation, the back-up DP control system and its associated reference systems are to be provided with at least a dedicated UPS. The UPSs are to be located and arranged such that no single fault, including fire or flood, in one compartment will interrupt the power supplied to the remaining control system and associated reference sensors.

2.5 Energy Storage Systems Supporting Dynamic Positioning Operations

In general, Lithium type battery systems can be used to power electrical loads such as large power propulsion and vessel's loads. Typical Energy Storage Systems (ESS) can be of the lithium battery or supercapacitor type. Lithium batteries and/or supercapacitors installations are to comply with the related Class specific Rules. Refer to Table 2.1 for reference.

2.6 Considerations When Installing Energy Storage Systems

When installing ESS supporting the DPS system, the following is to be considered in the design phase of the project:

- The ESS power is to be reserved for use by critical loads and not depleted by others.
- When the energy storage and conversion equipment is located downstream of the rectifiers that feed those loads, it is not to feed a distribution side fault and thus there is to be no change in the fault characteristics of the power plant. Coordination studies and short circuit test results are to be validated.

Note: When battery Energy Storage Systems are retrofitted to a vessel that has already undergone successful validation testing for fault ride through in closed bus ties configuration and the batteries are only installed downstream of thruster (or other) rectifiers (uncontrolled rectifiers), there will be no requirement to revalidate coordination studies or carry out additional short circuit testing because the batteries cannot feed a distribution side fault.

- The battery installation is to be distributed so that the concentration of flammable or explosive materials in any one space are reduced.
- Connecting the batteries at the thrusters' bus level rather than to the main distribution system's busbars may require installation of batteries at other more locations.
- The ESS system is to be sized and selected accordingly to operate and provide its functionality without damage, effect, or degradation throughout its service life.

Note: *The text* underlined *above is sourced from MTS document titled* "DP Vessel Design Philosophy Guidelines".

2.7 Fault Ride Through (FRT) Capability

Fault ride through (FRT) capability is to be provided for:

- Vessels with the **DPS-3** and **DPS-3 +** notations as per the section entitled *Vessels with* **DPS-2** *or* **DPS-3** *notations.*
- Vessels with the **EHS-E** and **EHS-P** notation as per Chap. 7, *Enhanced systems* and as per the section entitled *Vessels with* **DPS-2** *or* **DPS-3** *notations.*
- Fault ride through (FRT) capability is to be demonstrated first by computer modelling and simulations as per the section entitled *Computer modelling and simulations* below.
- Secondly by a physical live test, e.g., a short circuit test (or any other equivalent fault ride through live test) as per the section entitled *Live test requirements* below.
- In general, the power system attributes are to be proven by the combination of the simulation and testing process to demonstrate the FRT capability. The attributes are listed in 5.2 of the in Marine Technology Society (MTS) document *TECHOP_ ODP _09(D), May 2015.*
- Documentation supporting the computer modelling process and simulation results are to be provided to Class for review.
- Documentation supporting and demonstrating the physical live tests (e.g., test protocols, tests setups, methodologies, results, etc.) is to be provided to Class for review.
- Consider risks associated with the test. A risk assessment is to be conducted to identify if the live test is adequate.
- In case the risk associated with a live test is found to be high, virtual test(s) may be considered in lieu of live test(s), provided the power plant model has been validated through lower risk level live test(s).

Note: Vessels which cannot demonstrate fault ride through capability by testing (live test or virtual test), due to limitation of the design, should not operate their power plant in configurations using closed bus ties.

- Model files developed during the modelling, simulation and validation processes are to be available to Class upon request.
- Live tests are to be witnessed by the Class approved Surveyor.

The fault ride through is the capability of a vessel's DP system power plant and control systems (for a closed-bus operation) to recover from the effects of a fault anywhere on the main and distribution systems (e.g., on a main switchboard bus segment or on a major feeder circuit) without significant malfunction before the bus tie breaker opens to isolate the faulted side of the system during a "limited" time or voltage drop. "Limited" time is to be defined by the designer in order to make the system stable. ***Note:** In general, common points that span the redundancy groups have the potential to introduce fault propagation pathways. An example of compensating provisions to address such vulnerabilities is to demonstrate fault ride through capability of equipment that may be impacted by faults propagating through the common point.*

2.7.1 Pre-requisites for a Live FRT Capability Test

2.7.1.1 Computer Modelling and Simulations

The entire electrical network is to be modelled for a specific FRT capability analysis. Computer aid tools appropriate for the analysis are to be used. Several fault locations on the electrical network, defined configurations, fault types, and/or operational modes are to be selected in order to determine the worst fault case scenario(s). Failure mode simulations are to include short circuits, earth faults, arcing faults, crash synchronisation, inadvertent connection of a stopped generator, generator's pole slipping conditions. See also attributes to be proven as per above. The most practical and least risk fault condition (based on the simulation case scenarios' results) is to be selected to conduct the live test as required in the section entitled *Live test requirements* below.

2.7.1.2 Equipment Design

Equipment is to be designed and built to facilitate testing or be of a suitable design and construction so as to facilitate safe, effective verification by testing. Vendor electrical equipment's data/parameters as needed are to be available for the modelling process. All equipment being tested is to meet Original Equipment Manufacturer (OEM) specifications to carry out the tests that verify and validate the simulations.

2.7.1.3 Resources

Fault ride-through testing is to be carried out by entities/organisations with the required technical knowledge/knowhow, tools (modelling/simulations/data capture) technical and operational expertise and competent resources. Organisations of this type are also to have the test equipment, software and facilities to carry out the modelling and simulation work such as Real Time Digital Simulation or Hardware-in-the Loop testing.

2.7.1.4 Live Test Requirements

In general, the live test is to follow the recommendations as listed in the Marine Technology Society (MTS) *TECHOP ODP 09 (D) (May 2015), A Method for Proving the Fault Ride Through Capability of the Vessels with HV Power Plant*. The physical live test, e.g., a short circuit test (or any other equivalent fault ride through live test) is to be in accordance with a prescribed step by step test protocol provided by the power system designer, integrator, and vendors in order to validate the simulation/analysis results as stipulated in *Computer modelling and simulations* above. All equipment essential for the DP system is to have fault ride through capability, allowing for a short circuit condition to clear before under voltage protection is actuated. Low voltage transients during a short circuit condition are not to cause a motor starter to drop out, or other drives to fail. One successful fault ride through capability test is required to validate the modelling work and analysis selected for the worst-case scenario(s). If major changes to the electrical network topology and coordination protection schemes are made throughout the life of the vessel, then the FRT capability tests are to be revalidated. The physical live tests are to be performed under an Electrical Hazard Safety Programme. Successful test results, along with all deviations and findings are to be submitted to Class for verification.

2.8 DC Power Distribution Systems

For DP vessels designed with a DC power distribution topology, they are to comply with the Class mandated rules for direct current (DC) power distribution systems.

2.8.1 DC Control Power Supplies and Battery Systems

DC control power and battery charger systems are to be provided, as a minimum, in accordance with the overall redundancy concept. The design is to consider addition of non-critical redundancy to improve availability. For example, a second battery supply to allow battery maintenance. The output of DC systems supplying equipment intended to provide redundancy is not to be cross connected. Crossovers for maintenance are not to be provided in DP Class 3 vessels if this can lead to transfer of fault by fire or flooding. Control supplies are to have a primary supply from the appropriate part of the power distribution system in a manner that supports the redundancy concept and a backup power supply from the emergency power distribution systems. The risk of transfer of fault and hidden failures is to be reduced to a minimum. Control supplies for thrusters and generators are to be provided in a manner that makes each thruster or generator as independent as possible. Control systems for sensitive circuits are not to be shared with heavy energy consumers such as circuit breaker spring winders. Simultaneous operation of multiple spring winders may cause control systems to malfunction.

2.8.2 DC Power Generation and Distribution System

The generator sets are to be distributed in a minimum of two bus bars connected by bus tie breakers. Where the main source of electrical power is for propulsion of the vessel, the main DC bus is to be subdivided into at least two sections, which are to be connected by DC circuit breakers or other approved means. The electric power sources are to be connected to the DC bus directly or through power electronic converters (such as AC/DC rectifiers or DC/DC converters). The electrical power sources are to be protected against faults with protective devices such as DC circuit breakers, fault current limiting/bypass converters, solid state switches and fuses, or a combination. An ESS system connected to the DC busses can be considered as static spinning reserve and are to be able to handle peak shaving for variable loads. Their capacity is to be considered in the overall system calculation. The ESS system is to be controlled and interfaced to the system's energy management system (EMS). Generator speed and load curves are to be automatically controlled by EMS based on actual system load. The DC bus is to be properly sized to withstand the short circuit current available on the DC bus and arranged such that the temperature rise will not affect the normal operation of electrical devices connected to it. The DC bus tie breaker is to be an ultrafast acting device able to trip a short circuit condition at voltage levels such as 1,500 VDC to enable selectivity between functional groups. Coordination between the AC generator protection and rectifier module protection during a short-circuit condition on the DC bus is to be provided for DC distribution systems supplied by AC generators.

Chapter 3
Thruster System Design

Thrusters are to comply with the requirements of Part 4, this chapter of the Marine Vessel Rules, for vessels subject to ABS Classification, or as applicable. The thruster system is to be designed for continuous operation. When the main propulsion propellers and rudders are included under DP control, they are considered as thrusters and all relevant functional requirements of this chapter are to be applied. When the main steering system is included under DP control, the steering gear is to be designed for continuous operation.

3.1 Thruster Capacity

The thruster system is to provide adequate thrust in longitudinal and lateral directions and yawing moment for heading control.

- **Vessel with DPS-0 or DPS-1 notation.** The vessel is to have thrusters in number and of capacity sufficient to maintain position and heading (station keeping) under the specified maximum environmental conditions.
- **Vessel with DPS-1 + notation.** Transverse thrust force and yaw moment are not to be based on the propellers and rudders. The thruster configuration is to be arranged to provide the required transverse force and yaw moment for station keeping.
- **Vessel with DPS-2 or DPS-3 notation.** The vessel is to have thrusters in number and of capacity sufficient to maintain position and heading (station keeping), in the event of any single fault, under the specified maximum environmental conditions. This includes the failure of one or more thrusters.

F. Karkori, *Dynamic Positioning Systems*, Springer Series on Naval Architecture, Marine Engineering, Shipbuilding and Shipping 21,
https://doi.org/10.1007/978-3-031-59173-0_3

- **Vessel with DPS-2+ or DPS-3+ notation.** The vessel is to have thrusters in number and of capacity sufficient to maintain position and heading (station keeping), in the event of any single fault, under the specified maximum environmental conditions. This includes the failure of one or more thrusters. Transverse force and yaw moment are not to be based on the propellers and rudders. In post-failure conditions, the combination of propellers and rudders may be accepted as a side thruster for transverse force for back-up purposes only when the engine is continuously running in minimum RPM with the propeller pitch at zero during the DP operation.

Thruster installations are to be designed to minimise interference with other thrusters, hull and other surfaces. In the calculation of available thrust in the consequence analysis and station keeping capability analysis, the interference effect between thrusters and other effects that could reduce the available thrust is to be considered. A single fault in the thruster system, including pitch, azimuth or speed control, is not to result in unintended operation of pitch, speed and direction.

3.2 Thruster Configuration

The thruster arrangement is to provide adequate manoeuvrability under all operating conditions. The layout of the multiple thrusters is to consider the need for continued operation with the loss of the most effective or critical thruster(s), as well as the need to maintain position and or heading upon the onset of the worst-case failure, where multiple thrusters may be lost. (i.e., Worst Case Failure). A single fault in the thruster system is to be such that the thruster fails to a safe mode so that the vessel's position and heading are not affected. Fail to a safe mode could be a failure to zero thrust or motor stop.

3.3 Thruster Auxiliary System

Auxiliary systems whose functions have a direct effect on the thruster system, for example, lubrication oil, cooling water, control air and uninterrupted power supply systems, are to be provided for each thruster system in a manner that supports the worst-case failure design intent.

3.4 Thruster Control

Built-in features are to be provided to prevent overloading of the drive system. Manual override of control is required so that pitch, rpm, azimuth or other parameters of individual thrusters can be controlled and can be stopped if necessary. Thrusters in DP operation is to provide controllable thrust from zero load to full load in step-less increments. This can be achieved through control of the propeller pitch or the speed of the propeller, or other parameters. For **DPS-1**, **DPS-1 +** , **DPS-2**, **DPS-2 +** , **DPS-3** and **DPS-3 +** notations, the DP control system and manual position control systems (Chap. 4) are to be provided for the thruster system. In addition, a manual thruster control system is to be provided. For **DPS-0** notation, a manual position control system is to be provided.

3.4.1 Control Parameter

Command of thruster pitch, azimuth and rpm is to be provided by a closed-loop control system where feedback signals are compared with the sent commands. Sensors connected to the system are to provide independent signals representing each of the above three parameters. For fixed pitch propeller thrusters, two parameters of azimuth and rpm are to be represented through sensors. If the dynamic positioning system uses other parameters, such as thrust, amperage, and power, in the closed-loop control system, those parameters are to be monitored.

3.4.2 Manual Thruster Control

The manual thruster control system is to be independent of the DP control systems so that it will be operational if the automatic control systems fail. The system is to provide an effective means of individually controlling each thruster from main DP control station. The system is to provide an individual lever for each thruster and to be located at the main DP control station. If not practical to provide an individual lever for each thruster, sharing a lever between thrusters may be acceptable provided that following conditions are satisfied:

- For **DPS-0** and **DPS-1**, thruster control levers can provide controls of thrust in transverse and longitudinal directions and resulted yaw moment at same time.
- For **DPS-2** and **DPS-3**, thruster control levers can provide controls of thrust in transverse and longitudinal directions and resulted yaw moment at same time after any single fault. For **DPS-3**, single fault also includes a loss of a compartment due to fire or flood.

Any failure in the manual position control system is not to affect the capabilities of the manual thruster control system to individually control each thruster or related group of thrusters. Manual thruster control is to be available at all times, also during all failure conditions in the main DP control system. A single fault in the manual thruster control system is neither to cause significant increase in thrust output nor make the thruster rotate.

3.4.3 Emergency Stop

An emergency stop facility for each thruster is to be provided at the main DP control station. The emergency stop facility is to be independent of the DP control systems, manual position control system and manual thruster control system. The emergency stop facility is to be arranged to shut down each thruster individually. This emergency stop is to be arranged with separate cables for each thruster. Electrical cables potentially exposed to hydrocarbon fires in engine rooms and spaces where fuel oil is contained these cables are to be fire-resistant coated. An alarm is to be initiated upon loop failure (i.e., broken connections or short-circuit) in the emergency stop system. The emergency stop activation buttons are to be placed in a dedicated layout representing the thruster location and which is consistent with the vessel's axis and layout, or they may be arranged together with the corresponding thruster levers if these are arranged in accordance with the physical thruster layout. Where an accidental operation of the emergency stop buttons can occur, a protective cover is to be mounted. Emergency stops for thrusters are to be located within easy reach of the DP operator (DPO) e.g., within the bridge, at the main DP control station. Emergency stops for thrusters are to be laid out in a logical manner which reflects the position of the thruster in the vessel's hull. For equipment classes 2 and 3, the thruster emergency stop system is to have loop monitoring.

3.5 Thruster Monitoring and Alarm

Thruster monitoring is to be provided at the DP control station. This monitoring is to be continuously available by indicators/screens. The following parameters are to be monitored:

- RPM, pitch, azimuth.
- CPP hydraulic oil pressure and temperature.
- Thruster load and motor temperature.
- Thruster motor/semiconductor converter coolant leakage.
- Thruster motor/semiconductor converter temperature.
- Thruster motor power availability.
- Thruster motor short circuit.

- Lube oil pressure and temperature.
- Thruster operation (on-line/off-line).

Alarm systems for leakage, motor overload and high temperature, abnormal pressure are to be audible and visual and are to be installed on the DP control station. Their occurrence and reset status are to be recorded. Chapter 1, Table 1.2 summarises the parameters for monitoring, display and warning.

Chapter 4
Control System Design

In general, control and monitoring system components (alarms and instrumentation) for dynamic positioning systems of a vessel intended to be assigned a DPS notation are to comply with the requirements of the vessel's mandatory classification notations. ***Note:*** *good engineering practice promotes the concept of maintaining clear segregation between systems intended to provide (i) control; (ii) monitoring; and (iii) protection.*

4.1 DP Control Station

The general requirements for system arrangement for the different notations of **DPS-0**, **DPS-1**, **DPS-2**, and **DPS-3** are listed in Chap. 1, Table 1. In general, the following items are to be located at the DP control station:

- Operator stations for DP control and manual position control.
- User interface displays for required position reference systems.
- Manual thruster control levers.
- Thruster control mode change systems.
- Thruster emergency stops.
- Means to monitor thruster, main propeller and rudder performance (if relevant to DP) and manage unwanted thrust.
- Equipment for internal and external communication.

System components that do not require frequent or immediate operator attention may be installed in alternative locations. DP control systems and equipment are to meet the requirements of Class pertaining to the Class specific Rules for building and classing marine vessels.

F. Karkori, *Dynamic Positioning Systems*, Springer Series on Naval Architecture, Marine Engineering, Shipbuilding and Shipping 21,
https://doi.org/10.1007/978-3-031-59173-0_4

4.1.1 Arrangement of DP Control Station

The DP control station is to display information from the power system, thruster system and DP control system so that correct functioning of these systems can be maintained. For **DPS-3** notation, a backup DP control station is to be provided in a separate compartment located and arranged such that no single fault, including a fire or flood in one compartment, will render both the main and backup control system inoperable. This room is to be separated by an A-60 class division from the main DP control station and located with easy access from the main DP control station. The main and backup DP control stations are to be arranged so that the operator does not need to change position during operation and the operator has a view of the external surrounding area and all activities relevant to the DP operations. For the backup DP control station, alternatively, a closed-circuit TV (CCTV) system is acceptable for viewing the external surrounding area. If a view of the vessel's exterior limits and the surrounding area is not considered important for the intended operations, for example, for MODUs, the main activity after damage of the main DP control station is to cease the drilling operation, the arrangement for viewing the vessel's exterior may not be necessary for the DP backup control station.

4.1.2 Arrangement of DP Control Systems

DP control mode is to include control of position and heading. Set-points for control of position and heading are to be independently selectable.

- For the **DPS-0** notation, a manual position control system is to be fitted.
- For the **DPS-1** notation, a DP control system and a manual position control system are to be fitted. Transfer of control between the two systems is to be initiated manually.
- For the **DPS-2** notation, a DP control system with two DP control computers and a manual position control system are to be fitted. Transfer between two DP control computers is to be automatic and bump-less with no loss of position and/or heading.
- For **DPS-3** notation, a main DP control system with two DP control computers, a backup DP control system, and a manual position control system are to be fitted. The backup DP control system is to be located at the backup DP control station. The backup DP control system is to have an automatic position control mode and is to be interfaced with a position reference system, motion reference unit (MRU or equivalent) and a vessel heading sensor which are able to operate independently of the main DP control system. The transfer between the main DP control system and the backup DP control system is to be smooth with no loss of position and/or heading. Refer to the section entitled *Main/backup control station* for modes of transfer.

In general, each DP computer system is to be isolated from other on-board computer systems and communications systems for maintaining the integrity of the DP system and command interfaces. This isolation may be achieved using via hardware and/or software systems and physical separation of cabling and communication lines. Robustness of the isolation is to be verified by analysis and proven by testing. Specific safeguards are to be implemented to secure the integrity of the DP computer system and prevent the connection of unauthorised or unapproved devices or systems.

4.1.3 Arrangement of Control Panels

The operator is to have easy access to information on displays or by visual indication. Display systems and the DP control station are to be based on sound ergonomic principles which promote proper operation of the system. The effect of any action is to be immediately displayed, preferably with graphics. In addition to the DP system command signals, the feedback signals are to be displayed where applicable. The selection between operational modes (DP mode/manual position control mode/ manual thruster control and if part of the thruster system, propellers and rudders) is to be provided and easily operated. Indications of active mode and the operational status of different systems are to be displayed on the control panels. Indicators and controls are to be arranged in logical groups and be coordinated with the geometry of the vessel where applicable. Indication of the system in charge, and changes in operational mode, is to be clearly provided to the DP operator. Indications of changes in operational mode is to be visually indicated along with an audible alarm. Manual operating levers and other controls are to be placed in close proximity to the DP operator and intuitive in nature. Means for preventing inadvertent operation of vital or critical controls are to be considered such as recessed or covered switches and others. Protection of those buttons can be provided by a two-button arm/fire system, or by placing a passive cover over them. If an erroneous sequence of operations may lead to a critical situation or damage of equipment, appropriate interlocks are to be arranged. Controls and indicators placed in the DP control stations are to be lighted properly to permit use day and night. Lights for such purposes are to be easily adjustable.

4.1.4 Arrangement of Data Communication Networks

When the DP control system utilises a data communication network to distributed thruster control units, this network is to be separated from the communication network(s) for manual thruster controls and manual position control. The communication network for the DP control system is to be duplicated and each is to be routed as far apart from each other as practical for **DPS-2**. The network is to be duplicated and physically separated for **DPS-3**. The manual position control system is not to share the same communication network with the DP control system. DP related networks, including data communication networks, are likely to introduce common points that span the redundancy groups. Such common points have the potential to introduce fault propagation pathways which can defeat the redundancy concept. These common points are to be clearly identified in the DP FMEA and addressed appropriately (e.g., Single Failure Propagation Analysis concluding on the local effect and global effect, compensating provisions, or mitigating measures and efficacy of same to be proven by validation testing, etc.).

4.2 DP Control System

In general, the DP control system is to be arranged in a DP control station. This system is to be located so that the operator can easily maintain an overview of both the vessel and the surrounding area. Design is to be robust enough to reject erroneous signals that could affect control of position and/or heading. If all position reference systems fail the DP control system is to use the internal model to send control signals. The transition between the DP control modes is not to lead to a significant change in thrust output.

4.2.1 External Force

If there are external forces from other equipment that can affect the vessel positioning such as from pipe tensioners, hawser tension, or fire monitors; then a comprehensive system engineering approach is to be implemented for automatic or manual feed forward compensation. This is to consider the external sensor redundancy and failure modes (benign and aggressive failure modes). In addition to error checking functions provided by redundancy, the acceptable signal ranges are to be carefully bounded to be within a realistic range.

4.2.2 Vessels with DPS-2 Notation

DP control system is to consist of at least two independent computers systems so that, in case of any single failure, automatic position keeping and heading control ability will be maintained. Failure of common facilities, such as self-checking routines, alignment facilities, data transfer arrangements, and plant interfaces, is not to cause the failure of more than one computer system. An alarm is to be initiated if any computer fails or is not ready to take control. The DP control system is to perform self-monitoring and automatically transfer DP control after a detected failure in one control computer. The automatic control transfer from one control computer to another is to be bump-less with no loss of position and/or heading (station keeping). The cabling for the DP control computers and the thrusters is to be arranged such that under single fault conditions it will be possible to control sufficient thrusters capable of providing thrust to control surge, sway and yaw and to keep the vessel within the specified operating envelope. An alarm is to be initiated for any failure of a DP control system or a sensor and also for the cases where the sensor or control system is not activated or on standby. Self-monitoring and comparison between systems are to be provided and warnings are to be generated upon detection of an unexpected difference in thruster command, position or heading.

4.2.3 Vessels with DPS-3 Notation

In addition to the requirements for the **DPS-2** notation, the following requirements also apply:

- The computer systems described above do not include the backup computer system, thus, an additional DP control system is to be located in the backup DP control station and transfer of control to it is to be initiated manually from the backup DP control station.
- The backup DP control station is to have means to detect abnormality of the backup system continuously and to communicate its status to the main DP control station. An alarm is to be initiated if the backup DP control station fails or is not ready to take control.
- The backup DP control station is to be ready to operate and maintain the position once it assumes command. The switch-over of the control to the backup DP control system is to be operated from the backup DP control station and is to occur via manual transfer when the main DP control station is affected by failure or fire.
- The cabling for the DP control systems and the thrusters is to be arranged such that under single fault conditions, including loss of a compartment due to fire or flood, it will remain possible to control sufficient thrusters to keep the vessel within the specified operating envelope.

4.2.4 Vessels with DPS-1+, DPS-2+ or DPS-3+ Notation

For **DPS-1+**, **DPS-2+** or **DPS-3+** notation, the main DP control system is to include a common joystick for the manual control of thrusters with mode selection of automatic and manual heading control. This requirement is applicable to the backup DP-control system for **DPS-3+** notation.

4.3 Manual Position Control System

The manual position control system is to be independent of the DP control systems and is not to rely on common cabling. The system is to be arranged such that it will be operational if the DP control systems fail. The system is to provide one joystick for manual control of the vessel position and is to be provided with the arrangements for automatic heading control.

4.4 Control Mode Selection

4.4.1 Manual/DP Control Modes

A simple device is to be provided in the DP control station for the selection of the thruster control modes (i.e., manual thruster control, Manual Position Control and DP control). The device is to be designed so that it is always possible to select manual thruster controls after any single fault in the DP control mode. Thrusters within the DP control system may also be individually de-selected from DP control to manual thruster control for service and vessel specific operations.

4.4.2 Main/Backup Control Station

For **DPS-3** notation, the mode selector between main DP control station and backup DP control station is to comply with redundancy requirements. The transfer to the backup DP control station is to be fail-safe, so that if the main DP control station is damaged in any way, transfer of control can still be initiated and assumed at the backup DP control station. Transfer of control to the backup DP control station is to be performed manually, such that inadvertent control transfer to an unattended station is avoided.

4.4.3 Position Reference System and Environment Sensor

The Position reference systems are to be in compliance with the relevant requirements for the mandatory classification notations of the vessel for electrical, mechanical, and hydraulic components and subsystems. Accuracy and limitations of the position references used are to be adequate for the specific task in which the vessel is engaged. The design of position reference systems and environment sensors with respect to power, signal transmission, and interfaces is to follow the system redundancy requirements. Sensors and/or position reference systems may be shared with other systems provided failure in any of the other systems cannot cause failure or loss of performance of the DP control system.

4.4.3.1 Motion Reference Unit (MRU)

Where position reference systems are dependent on correction of the measurements for roll and pitch noise, motion reference unit (MRU) or equivalent is to be provided.

4.4.3.2 Diversity

Where more than two positioning reference systems are required, at least two are to be based on different measurement techniques and each of the two are to be independent with respect to signal transmission and interfaces. Special consideration may be given where the use of different measuring principles is not practicable.

4.4.3.3 Loss of Inputs of Position Reference Systems

Loss of one or multiple position reference system inputs or sensor inputs are not to lead to significant change in thrust output. In general, upon recovery of the inputs the last position or heading set point before the loss is not to be automatically applied unless this feature is required for specific functionality that has been developed by the DP Control System vendor. Such functionality is to be subjected to a comprehensive system engineering approach (e.g., FMEA) and validated by testing. The DP operator may choose any other set point if the set point suggested by the control system is significantly different from the actual vessel position and/or heading.

4.4.3.4 Alarms

Alarms for sensor malfunctions are to be included in monitoring of positioning reference systems and environment sensors. An alarm is to be initiated on detection of a sensor failure, even if the sensor is in a standby or offline use at the time of failure.

4.4.3.5 Power Source

For the **DPS-1**, **DPS-2** or **DPS-3** notation, the power source to the position reference systems and environmental sensors are to be supplied from UPSs. Following components of the position reference systems, where applicable, may not need to be powered by UPS.

- Transducer hoist systems.
- Taut wire derrick control system.
- Heave/tension compensation (taut wire).
- Part of gangway system.

4.4.4 Vessels with DPS-0 or DPS-1 Notation

For the **DPS-0** notation, a position reference system, a wind sensor and a vessel heading sensor are to be fitted. For the **DPS-1** notation, they are to be provided in duplicate. Special considerations may be given where the use of different measuring principles is not practicable. The two independent position reference systems are to be installed and be simultaneously available to the DP control system during operation. Where user interface computers and displays are required for position reference systems, they are to be independent of the DP control system. The user interface display is to be placed at the DP control station and easily readable by the DP operator. Power supply to the position reference systems is to be from the corresponding UPS connected to its associated sensors (except for notation **DPS-0**).

4.4.5 Vessels with DPS-2 Notation

In addition to the requirements for the **DPS-1** notation, a third independent position reference system, a third wind sensor and a third vessel heading sensor are to be provided. A single fault is not to affect simultaneously more than one position reference system (i.e., no common mode failures). Where user interface computers and displays are required for position reference systems at least two of them are to be independent of the DP control system. The user interface displays are to be placed at the main DP control station, and they are to be easily readable by the DP operator. UPSs are to be provided for the power supply of the position reference systems and associated sensors, and they are to be in accordance with the overall redundancy requirements.

4.4.6 Vessels with DPS-3 Notation

In addition to the requirements for **DPS-2** notation, the third wind sensor, third vessel heading sensor and the third independent position reference system are to be directly connected to the backup DP control station and with their signals repeated to the main DP control station with appropriate signal isolation. The sensor receivers connected directly to the backup DP control system are in general to be installed in the same compartment as the backup DP control system. For sensors that cannot be located in the same compartment as the backup DP control station room (e.g., wind sensor, GPS antenna, taut-wire or hydro acoustic transducer), redundancy is to be provided as far as possible by separation and physical distance. The independence between main and backup DP control systems is to be provided. Signal isolation of position reference system and sensors is to be provided, such that failures cannot propagate from one system to the other system.

4.4.7 Signal Processing

Position reference systems are to be capable of providing new position data with an update rate and accuracy suitable for the intended automatic DP operations. The individual position reference system is to be self-operational, and not dependent on the DP control system for operation. For satellite-based systems, interface and necessary equipment for receiving differential correction is to be installed. Where several position reference systems are required, the DP control computers are to use signal processing techniques to validate the data received. The reference position of each system is to be available at the operator's request. Position reference systems that rely on motion reference compensation are to be provided with a means of selecting available motion reference units for use. When several systems are combined, the resulting value used in further processing is not to change suddenly or significantly if one or more systems are selected or deselected. Failures in a positioning reference system that might affect positioning or redundancy adversely are to initiate an alarm. If a positioning reference system can stall or in other ways produce corrupt or out of range data, a method is to be provided to enable rejection of the data and an alarm is to be initiated.

4.5 Consequence Analysis (for DPS-2 and DPS-3)

For a vessel with **DPS-2** or **DPS-3** notation, the DP control system is to incorporate a consequence analyser that monitors the composite thrust vector necessary to maintain position and heading under the prevailing environmental conditions in the event of predefined worst-case failure. Additionally, the consequence analyser is to be able to perform calculations to verify that in the event of a single fault there will be sufficient thrust available to maintain position and heading. Failure conditions may also be simulated under hypothetical conditions (e.g., loss of thruster, generator, auxiliary systems, etc.). This simulation is to be clearly identified and is not to affect the actual control system in any fashion. The consequence analysis is to be automatically running at pre-set intervals. The calculation ongoing is to be displayed. The analyser is not to be disabled while in **DPS-2** or **DPS-3** DP operations. The sampling or filtering of the consequence analysis is to be such that a useful warning can be given in sufficient time but without initiating numerous nuisance alarms which will bring the credibility of any warnings into doubt. When the consequence analyser predicts a failure mode that would cause a loss of position in the prevailing weather condition, and alarm is to be initiated and acknowledged by the DP operator.

4.6 Display and Monitoring

The DP control station is to receive alarms and warnings reflecting the status of the DP system. If the vessel exceeds preset position and heading limits an alarm is to be initiated. If the alarms in the DP control station are slave signals of other alarm systems, there are to be a local acknowledgement and silencing device. The silencing device is not to inhibit new alarms. The display unit is to present a position plot including the location of the vessel relative to the reference sources. The plot may be vessel relative, or a true motion presentation. For DP control systems designed with redundancy, there is to be at least two independent position displays. If the display is used for presentation of warnings or alarms, they are to have priority over other information and they are not to be inhibited by other data currently being displayed. The critical DP alarms and displays are to be available in **DPS-3** backup DP control station. Non-critical DP alarms (e.g., grouping of alarms, logging on by authorised crew only to the alarm monitoring system, etc.) are to be of non-intrusive type announcements for the DP operator, since in a fire or loss of compartment situations, many failed electrical systems, thrusters and other alarms may be initiated simultaneously.

Chapter 5
Auxiliary System Design

The design of the components of auxiliary systems is to be in compliance with the requirements of the vessel's mandatory classification notations. For **DPS-2** and **DPS-3** notations, the auxiliary systems that are part of the DP system are to be arranged in accordance with the redundancy requirements given in Chap. 2. A single failure effect analysis for auxiliary systems is to be included in the DP system FMEA. The following items are to be considered where applicable:

- Fuel oil system.
- Lubricating oil system.
- Cooling water system.
- Compressed air system.
- Hydraulic system.
- Pneumatic system, etc.
- Ventilation/HVAC system.
- Piping system equipment (e.g., purifier, heat exchanger, transfer pump, etc.).

5.1 Fuel Oil

The fuel system is to be arranged to follow the redundancy concept. The duty pump is to be powered from within the same redundancy group as the component or system it serves. The effect of pump failure is not to exceed the WCFDI. If the fuel system requires heating, then the heating system is to be designed with the appropriate level of redundancy. The design of the fuel system is to facilitate isolation of services between DP operation and industrial functions if applicable. Actuators for quick closing valves are to be installed on a per engine basis and hence, any remote-control system is to fail safe with respect to station keeping. Fuel strainers and filters are to be arranged to facilitate changes without taking equipment out of service. For **DPS-3**, a minimum of one service tank is to be provided for each redundant group.

F. Karkori, *Dynamic Positioning Systems*, Springer Series on Naval Architecture, Marine Engineering, Shipbuilding and Shipping 21,
https://doi.org/10.1007/978-3-031-59173-0_5

The service tanks are to be in separate compartments with A-60 partitions following overall split redundancy concept. The valves in the crossover facilities, if arranged, are to be located as close as possible to the bulkhead and operable from both sides.

5.2 Cooling Water

The cooling system is to be arranged to follow the redundancy concept. The duty pump is to be powered from within the same redundancy group as the component or system it serves. The effect of pump failure is not to exceed the WCFDI. For twin screw vessels where cooling pumps are engine driven, a duplicate spare pump carried onboard, in lieu of the standby pump, is acceptable, as long as loss of pump would not exceed the WCFDI. Engine room sea water cooling systems could be incorporated into the thruster seawater systems provided the redundancy concept is not violated. In consideration of the severe consequences arising from the loss of water or gas accumulation, freshwater cooling systems are to be arranged with full separation between systems providing required redundancy. For a **DPS-2** notation, the pipe can be considered not to fail. The fail-safe condition for valves in the freshwater system is to fail as is. The fail-safe condition for temperature regulating valves is to be fail open. The display of running pumps is to be provided and an alarm is to be initiated for abnormal stop and/or change over to a standby pump.

5.3 Compressed Air

Compressed air systems for DP related functions are to follow the redundancy concept. Compressed air for starting engines is to be independent to the maximum extent feasible. Control air and starting air may be taken from the same source provided any pressure drops associated with starting air do not affect the control function. Loss of air supply to the thrusters is to be alarmed and is to have no effect on thruster operation.

5.4 Lubrication Oil Systems

Lube oil systems for engines are to be associated with one engine only. Where common facilities are provided, such as, facilities for storage, changing and disposing of oil, means are to be provided to prevent inadvertent cross connections between engines which could lead to one engine sump being emptied and the other overfilled.

5.5 HVAC and Ventilation

Ventilation and HVAC for spaces containing equipment essential to DP are to be arranged to comply with redundancy so that acceptable temperature can be maintained after any single fault in active components and ventilation damper actuation energy source. This requirement also applies to switchboard rooms and instrument rooms containing components that are parts of the DP system.

5.6 Piping

Crossover pipes are acceptable, except in ventilation ducts, provided the cross over pipes can be closed at both sides of separating bulkheads. Crossover valves, where fitted between independent systems to facilitate maintenance, are to be provided with local and remote monitoring to indicate open/closed status.

5.7 Pneumatic Systems

Pneumatic systems are to be designed according to required redundancy in view of the risk of leakage.

5.8 Power Supply to Auxiliary Systems

Power for auxiliary systems associated with DP systems is to be taken from within the redundancy group. Auxiliaries for thruster systems such as cooling water pumps and fans are to be powered from the same redundancy group as that providing the drives.

Chapter 6
DP System Initial Testing Regime

The components and subsystems associated with the dynamic positioning system are to be tested according to the vessel's mandatory classification notations, where applicable, prior to testing of the dynamic positioning system. Once completed, the dynamic positioning system is to be subjected to final tests. These tests are to prove essential features of the DP system and its design redundancy. Objectives of the test are to verify the intent of the DP vessels redundancy design (worst case failure design intent) and to validate the same by testing. In order to achieve this objective, attributes of Performance, Protection and Detection on which the redundancy concept depends upon will need to be validated by testing. A DP test programme is to be developed to contain test procedures and acceptance criteria and identify the configurations for all DP related systems for test. The DP test programme is to include procedures for DP system performance testing and procedures for FMEA proving trials for **DPS-2** and **DPS-3**, as applicable. When deemed necessary by the attending Class approved Surveyor additional tests to those specified by the DP test programme may be required. The DP test programme is to be kept onboard the vessel. Any changes to the DP system other than like-for-like replacement of equipment are to be notified to Class and included in a revision of the DP test programme.

6.1 DP System Performance Test

The performance testing of the DP system is to be carried out to the satisfaction of the attending Class approved Surveyor in accordance with a DP system performance testing procedure which includes but is not limited to the following:

F. Karkori, *Dynamic Positioning Systems*, Springer Series on Naval Architecture, Marine Engineering, Shipbuilding and Shipping 21,
https://doi.org/10.1007/978-3-031-59173-0_6

6.1.1 Uninterruptible Power Supply Systems (UPS)

The UPSs are to be operated and confirmed to be functioning satisfactorily. The UPSs are to be operated without the normal main power input for 30 min to confirm that the batteries are capable of supplying the output power with full functioning of the DP control system (DPCS). The alarms for the UPSs are to be tested. ***Note:*** *UPSs play a prominent role in the integrity of the DP system and are likely to be present in systems other than the DPCS UPSs, (e.g., UPSs for control power of switchboard, mission critical UPS's with interfaces to the DP System* etc.*).* ***Note:*** *Fault ride through capability and fault tolerance capability may depend on UPS's. It is important that such dependencies are clearly identified and reflected in the Failure Modes and Effects Analysis. By-pass and synchronisation functionality, if present is to be comprehensively analysed and tested to verify that malfunction/mal-operation does not defeat the redundancy concept.*

6.1.2 Position Reference Systems and Sensors

All position reference systems, sensors, MRUs, and vessel heading sensor(s) for the DP systems are to be tested and confirmed to be functioning satisfactorily. The alarm systems are to be checked by simulating failures of the sensors and the reference systems.

6.1.3 Manual Position Control System

For a **DPS-1**, **DPS-2** or **DPS-3** system, the operation of the manual position control system using one joystick at DP control station and supplemented by an automatic heading control, is to be confirmed to be functioning satisfactorily.

6.1.4 Manual Thruster Control System

The operation of the manual thruster control system using individual levers, at the DP control station in accordance with Chap. 3 is to be confirmed to be functioning satisfactorily.

6.1.5 Thruster Emergency Stop

The operation of thruster emergency stop, and monitoring alarms, including loop monitoring, is to be tested.

6.1.6 DP Control System

For a **DPS-1** system, the operation of the automatic control system and a manual position control system including manual transfer of control between the two systems is to be confirmed to be functioning satisfactorily. For a **DPS-2** system, the operation of the DP control system and a manual position control system including automatic transfer of one DP control computer to another upon failure is to be confirmed to be functioning satisfactorily. Upon failure of the DP control system, it is to be verified that the manual position control is possible. For a **DPS-3** system, the operation of three DP control computers and a manual position control system including main DP control system automatic transfer of one DP control computer to another upon failure is to be confirmed to be functioning satisfactorily. Manual transfer of control is to be verified to be possible at the backup DP control system located in the back-up control station. Upon failure of the main DP control system, it is to be verified that the manual position control is possible. Performance of the position reference systems in DP mode is to be tested for all possible combinations of position reference systems (PRS), and on each PRS as a single system. The main DP control system in a **DPS-2** or **DPS-3** system is to perform self-check routines which are to automatically changeover to a standby system when critical failure conditions are detected. An alarm is to be initiated in case of failure.

6.1.7 Control and Alarms

Functions of thruster controls and related alarms are to be tested and to be confirmed functioning satisfactorily. Change of thruster controls between DP control system, manual position control system and manual thruster control system is to be tested and to be confirmed functioning satisfactorily.

6.1.8 Standby Changeover

Changeover to standby pumps (e.g., sea water, fresh water, hydraulic) is to be tested if these pumps provide essential redundancy, (e.g., standby seawater cooling pump which is required to start on failure of the running pump). Any other standby changeovers that are included in the Operations Manual are to be tested.

6.1.9 Protection Function

The functions of protection equipment are to be tested if they are designed to provide essential redundancy of the DP system.

6.1.10 Communication

A means of voice communication between the main DP control station, and the thruster room(s) is to be tested and confirmed to be functioning satisfactorily. A means of voice communication between the main DP control station, the engine control position and any operational control centres associated with DP is to be tested and confirmed to be functioning satisfactorily. Refer to Chap. 1 for continuity of power supply for essential communication.

6.1.11 DP Endurance Test

A DP endurance test is to be carried out for at least 6 h with the complete DP system in operation. It can be combined with the test of DP control systems. The failures related to DP system performance are to be recorded and analysed. Where possible the test environment is to reflect the limiting design operating conditions.

6.2 FMEA Proving Trial for DPS-2 or DPS-3

For a **DPS-2** or **DPS-3** system, FMEA proving tests are to be carried out to the satisfaction of the attending Class approved Surveyor. The tests are to confirm the findings from FMEA analysis that the vessel is able to maintain position and heading after the loss of redundant groups identified by the FMEA, as well as clearly identify and prove the vessel's Worst-Case Failure and validate that the Worst-Case Failure Design Intent has been achieved. The FMEA test procedures are to be developed in the FMEA analysis covered in Chap. 1 of this book. The vessel's DP system FMEA is to have analysed and tested all the configurations that the vessel is intended to be operated in over its life cycle (e.g., closed bus configurations, and FRT tests). The test procedures are to be based on the simulation of failures and is to be carried out under as realistic conditions as practicable. Validation testing is to be carried out to demonstrate that the common point introduced by the closing of the bus ties does not provide a fault propagation pathway, and equipment related to DP Station keeping (including auxiliary systems and supporting systems) demonstrate voltage dip fault ride through capability. The schedule of these tests is to be designed to demonstrate the level of redundancy established in the FMEA. Where practicable, the test environment is to reflect the limiting design operating conditions. Closing out of action items from the FMEA, and proving trials are to be well documented and auditable. After completion of the DP proving trials, the final version of the DP FMEA and DP proving trial report, including final analysis/conclusions based on actual results from DP testing, are to be submitted.

Chapter 7
Enhanced System Notations

At the Vessel Owner's request, an Enhanced System notation (**EHS**), as a supplement to a **DPS** series notation, may be assigned to a **DPS-2** or **DPS-3** vessel. The main objective of the enhanced system notation is to improve reliability, operability and maintainability of the DP vessel. The enhanced system notations mainly emphasise the following properties of the DP system:

(1) Robust design of power plant and thruster system.
(2) Enhanced protection systems for generators and thrusters: Failure detection and discrimination of failed components before a full or partial black-out situation occurs.
(3) Automatic blackout recovery.
(4) Enhanced position reference system and sensor system for increased availability and reliability.
(5) Fire and flood protection for higher risk areas.

The notation includes provisions for standby start, closed bus operation and transferable generators. These are beneficial to the overall environment, operational flexibility and system maintainability. Four separate enhanced notations are provided as follows:

- Enhanced electrical system (**EHS-E**): This notation covers the requirements for the electrical and power management systems that are beyond those for the **DPS** series notations.
- Enhanced power and thruster system (**EHS-P**): This notation covers the requirements for the power system and thrusters that are beyond those for the **DPS** series notations.
- Enhanced control system (**EHS-C**): This notation covers the requirements on the DP control systems including control computers, position reference systems and sensors, which are beyond the minimum requirements for **DPS** series notations.
- Fire and flood protection system (**EHS-F**): This notation covers the requirements for fire and flood protection considering the risk level of the areas. This is a

F. Karkori, *Dynamic Positioning Systems*, Springer Series on Naval Architecture, Marine Engineering, Shipbuilding and Shipping 21,
https://doi.org/10.1007/978-3-031-59173-0_7

supplement for a **DPS-2** system. It is not necessary for a **DPS-3** system, since a **DPS-3** system has higher requirement in this regard.

The separate enhanced system notations provide the Owner with the flexibility of selecting an individual **EHS** notation or combined **EHS** notations that best fit the design intent. To be granted an **EHS** series notation, the requirements given in this chapter are to be met in addition to the requirements for **DPS** series notations.

7.1 Enhanced System for DPS-2 and DPS-3

A vessel with **DPS-2** or **DPS-3** notation may be assigned enhanced system notations if the system meets the additional requirements given in this section. For a vessel with a **DPS-2** notation, the Enhanced Power and Thruster System Notation (**EHS-P**), the Enhanced Control System Notation (**EHS-C**), Fire and Flood Protection (**EHS-F**) Notation or any combination, such as **EHS-PC**, **EHS-PF**, **EHS-PCF**, may be assigned. For a **DPS-3** vessel, **EHS-F** is not necessary since **DPS-3** has a higher fire protection requirement.

7.2 Enhanced Electrical System (EHS-E)

7.2.1 System Design

The design of power plant and thruster systems are to adhere to clearly defined worst case failure design intent. The design is to include fault ride through capabilities and functions for automatic quick blackout recovery. The design for transferable thrusters and generators is to be fault tolerant and fault resistant, which means that a single fault is not to cause failure of more than one redundancy group and is not to cause total black-out. A strategy for detecting hidden failure that could defeat the DP redundancy concept is to be developed. The strategy is to include means to effectively test protection systems and standby redundancy. Such systems are to be designed so that they can be tested on the Class approved Surveyor's request.

7.2.2 Power System

Power generation. Each generator set is to be autonomous in terms of control and automation functions. Common cause failures associated with load sharing, interlocks, permissive and others are to be avoided by distributing control and monitoring functions to local systems responsible for each item of main machinery.

Power distribution. With Enhanced Generator Protection (EGP), described below, information about bus load sharing conditions is to be directly derived from the bus bar. Generators are to be capable of droop mode operation, with any central communication infrastructure minimised.

(1) Load sharing. The load sharing system is to be designed so that the system can be tested, both for correct sharing and consequences of failure. The design is to incorporate the possibility for testing that critical failures are detected by the control system. The failure of the power management system is not to result in the failure of the load sharing system in any load sharing control modes.

- *Active (kW) load sharing.* If a scheme for isochronous load sharing is implemented, the generator speed controllers are to automatically switch to load sharing in droop mode upon failure of isochronous load sharing. Means to test the transfer conveniently are to be provided.
- *Reactive (kVAR) load sharing.* The generators are to be able to operate in voltage droop mode, without the need for reactive load sharing communication between the AVRs. If differential compensation between voltage regulators is implemented, the system is to automatically switch to voltage droop control mode operation if a failure in the reactive load sharing lines occurs. Means to test the transfer conveniently are to be provided.

(2) Switchboards. Switchboards are to be protected against potential arc damage either by arc detection, by insulated bus bars, or by any other means.

(3) Circuit Breakers. A minimum of two bus-tie breakers, one on each side, between switchboards is to be provided, with monitoring and control contacts from each. The Power Management System and generator incomer circuit breaker cubicle are to provide a visible indication of generator breaker status (open/closed). The Power Management System is to alarm if the status of the circuit breaker feedback differs from the issued command.

Power system protection. System design. Controls and protection systems are to be designed to facilitate local monitoring, controllability, fault detection, and acting in a timely manner to prevent fault propagating to other sections. A zone or segmented concept is to be implemented with error checking such as bus bar differential protection or equivalent. The generator protection system is to be designed so that it is possible to verify protection settings and test black-out recovery. Additionally, alarms and detection systems are to be capable of being tested after system delivery and be available for demonstration at the request of the Class approved Surveyor.

(1) Enhanced generator protection. In addition to the generator protection requirements for a vessel's mandatory notations, the following functions, as a minimum, are to be provided.

  (a) Identification of faults based on continuous monitoring of generator voltage and speed characteristics.
  (b) Monitor generator behaviour over time and reveal anomalous behaviour, and act before system performance is affected, or critical failures occur.
  (c) Trigger alarms if anomalies have continued beyond the pre-set intervals set by failure analysis, indicating that maintenance is required.
  (d) Identification of a faulty generator without relying on information from other generators.

EGP is to detect and alarm at least the following anomalies included but not limited to:

(a) Excess and insufficient fuel.
(b) Over and under-excitation.
(c) Generator instability or hunting.
(d) Loss of exciter current.
(e) Active and reactive power sharing imbalance.

All failures identified are to initiate an alarm or trip of the faulty generator as appropriate. The system is to be designed so that these features can be tested on the request of the Class approved Surveyor. If a generator is faulty, its circuit breaker is to automatically open, and the generator is to trip. In the event the breaker does not open, this is to be detected and the bus-tie breakers closest to the generator are to open automatically.

(2) Closed bus operation. If a failure is detected and an action is taken which fails to remove the failure, the protection system is to further isolate the fault by automatically isolating the affected switchboard. Bus frequency and voltage is to be monitored for operation within normal limits. If deviation occurs outside these limits for a pre-determined time interval, tie breakers are to automatically open, converting to a split bus system. An alarm is to be initiated.
(3) Blackout prevention system. Power consumers are to include intelligent blackout prevention that is not dependent on the Power Management System, including:

  (a) Dynamic limiting of power of the propulsion (and drilling services if applicable) according to available power.
  (b) Response time quick enough to avoid damage to the generators.

7.2.3 Thruster System

Thruster drives are to have short circuit, earth fault, thermal, over current and over-voltage protection. Motor starting and stopping controls are also to be provided near the drive. Each thruster drive control system is to have all the intelligence necessary to start the thruster and make it ready for DP control. Interlocks are to be provided to prevent a drive resetting itself under a short circuit condition. Thruster drives are to be able to start the thruster even when the propeller is turning due to inflow. ***Note***: *Thruster drives could be equipped with dynamic breaking resistors to permit more robust control of the drives operating point during various fault conditions.*

Interfaces. Where power management functions are integrated in electrical drives, the interface between the different power management functions is to be analysed in the FMEA. Where both the Power Management System and the local electrical drives can limit and control power to the thrusters, care is to be taken so they do not interfere with each other and cause a control oscillation between the two systems.

Fast phase back. To prevent power plant instability, variable frequency drives for the thrusters are to facilitate fast phase back; and in the case of drilling operations, phase back of the drilling system is also to be implemented. Care is to be taken so these two systems do not interfere with each other and cause a control oscillation. The criteria and order that equipment is phased back in (drilling equipment or thrusters) is to be described and added to the Operation Manual and be known to the DP operator and Class. Individual thruster drives are to directly sense under frequency in the system and cut back thrust.

Fault ride through capability. All equipment essential for the dynamically positioning system is to have fault ride through capability, allowing for a short circuit condition to clear before under voltage protection is actuated or the effects cause DP related consumers to malfunction. Low voltage transients during a short circuit condition are not to cause the motor starter to drop out, or other drives to fail.

7.2.4 Blackout Recovery

The power management system is to be able to re-establish the functioning of the power plant automatically, without any human intervention, within 60 s after blackout, generators and thrusters for DP are also to be available within 60 s. The system is to be able to re-establish the needed power for maintaining position and heading control within the specified envelope. The power management system is to be able to start all available generators after a blackout and to connect them to the electrical network.

Inrush currents. High inrush currents that may damage equipment and trip feeder breakers are to be avoided. This could be done by a pre-magnetising function for high voltage transformers or other equally effective means. If a pre-magnetising function is utilised and faults are detected during the pre-magnetising sequence, the connection of the specific piece of equipment experiencing a fault is to be aborted and an alarm is to be sent to the Power Management System.

Power distribution and management. To avoid problems with synchronisation, detection of a dead bus, defined here as a situation where the bus is de-energised, is to be secure, and in the form of a redundant detection system (e.g., dead bus relay and bus voltage transducer). A dead bus closing facility is to be implemented for overriding the synchronising function if the bus is dead. Two generators are not to be able to connect to the dead bus at the same time.

Thruster system. All thrusters are to be ready for operation without manual intervention. The thruster drives are to have their own local control system monitoring the condition of the main power system and reconnect as soon as the main power system is ready for the operation of the thrusters.

7.2.5 Uninterruptable Power Supply Systems (UPS)

If a system has independent subsystems for automation, control and supervisory systems, these are to be supplied by a dedicated UPS. The unit is to be of the online type as defined by Class, that is, there is to be no transfer time if a power failure occurs. The UPS are to have sufficient capacity to allow operations to be brought to a safe stop or discontinued with a minimum capacity of 30 min as required in Chap. 2.

7.3 Enhanced Power and Thruster System (EHS-P)

7.3.1 Systems Design

The system is also to comply with **EHS-E** requirements as prescribed in the section entitled *Enhanced Electrical System* (**EHS-E**) above. Additionally, the generator sets, and thruster sets are to be autonomous in the provision of auxiliary support services and control functions. Redundancy groups are to be clearly defined and well separated. Divisions are to be appropriately maintained throughout the design.

7.3.2 Power System

Power Generation. Each generator set is to be autonomous in terms of critical auxiliary systems (refer to *Auxiliaries* below), control and automation functions.

7.3.3 Thruster System

Autonomy. Each thruster set configuration is to be autonomous and operate independently from equipment that is not directly associated with its operation. Thruster seawater cooling system can follow the group redundancy concept with one duty and one standby pump. Engine room sea water cooling systems can be incorporated into the thruster seawater systems provided the redundancy concept (within the same redundancy group) is not violated. Fresh water-cooling system is per thruster. Cooling water for the thruster can also be used for auxiliary systems for that thruster.

7.3.4 Auxiliaries

Each thruster and generator set are to be autonomous in terms of auxiliary systems that are directly connected to it so that a single fault in the auxiliary systems is not to result in failure of more than one thruster/generator set. Separate fuel oil service tank for each generator is to be provided. These are to be located within their compartmental redundancy group. Fuel water content monitoring with remote alarms is to be installed either at the input to the service tanks or at the output of a purifier, whichever is more practical. Purification and transfer to the service tanks are to be arranged such that a single fault in the purification and transfer systems is not to result in failure exceeding the WCFDI. Fixed piping can be shared between generator/thruster sets in the same redundancy group. Separate freshwater cooling system for direct cooling of each engine is to be provided. External sea water or freshwater cooling systems can be on a per redundancy group basis, but in all cases, are to be arranged such that a single fault in external water-cooling systems is not to result in failure exceeding the WCFDI. External water-cooling systems for engine rooms could be incorporated into thruster external water-cooling systems provided the redundancy concept is not violated. Compressed air for starting engines is to be arranged such that a failure of one air compressor or air receiver is not to result in failure of more than one generator set. For standby units, on which the redundancy depends, the readiness of auxiliary systems is to be maintained and communicated to the DP-control systems. Units, which are intended to provide standby redundancy for each other, are not to share auxiliary systems, control power or starting facilitates.

7.4 Control System (EHS-C)

7.4.1 DP Control Computers

DP control station and DP controller. Three DP control computers (redundant main DP control system and a backup DP control system) each with their own independent power supplies are required. For **DPS-2** vessels, the manual position control system can be replaced with an additional DP control system (independent of the two DP control computers required by the basic **DPS-2** notation). The additional DP control system is to meet all of the requirements for the independent joystick control system while also providing DP control capabilities. Independence of the controller, workstation and network is to be maintained.

DP data logger. A data logger with easy playback facilities is to be implemented. The data logger is to stamp all data with time and date. Data logged are to include manual and automatic input to the DP control system and operational data of all DP relevant systems. Data recorded are to include status data and behaviour data of generators and thrusters. The data are to remain stored for at least 7 days, be easily accessible to the operator and be available for upload to offline storage media. Time and date are to be synchronised to a common reference. The data logger is to be able to log the main DP control system and the backup DP control system or additional DP control system (for **DPS-2**). The failure of the data logger is not to cause the failure of the two systems.

Capability plot. Online capability plots are to be updated using the tuned models on sea trials and include the simulation of the most relevant failure modes. Online capability plots are to be verified for accuracy during DP trials as possible.

Uninterruptable power supplies system (UPS). The power source of all control system components is to be supplied from a dedicated UPS. The UPSs distribution panel and fuse protection for the DP control systems are to be designed to isolate and clear faults individually.

7.4.2 Position Reference Systems

Four sets of position reference systems are to be implemented. The systems are to be designed so that three of the position reference systems are available at any given time and location considering blocking, noise, etc. One set of position reference systems is to be directly connected to the backup DP control system or additional DP control system (for **DPS-2**). The main DP control system is to be able to read the information from that system and the failure of that position reference system is not to cause the failure of both DP control systems. The position reference systems chosen are to reflect the planned operating conditions such as deep water (normally deeper than 500 m (1,640 feet), open water or close proximity to other units. In deep water (deeper than 500 m (1,640 feet) at least one of the position reference systems is to be an acoustic reference system. UPS supplies to position reference systems are to be sourced to prevent multiple sensor loss of the same sensor types.

Position reference system differentiation. No less than two of the position reference systems are to be a Global Navigation Satellite System (GNSS). At least one of these is to use GPS, GLONASS, or equivalent. At least one of the GNSS systems is to be a dual frequency receiver. The DP control system is to alarm any unrealistic GPS jumps in the signal or rapid drift.

Error checking. A system for error checking and weighting of the position references is to be provided.

7.4.3 DP Sensors

Four Motion Reference Units (MRUs) and four vessel heading sensors are to be implemented, with at least one based on different operating principles to the others a different manufacturer and working on different principles. A minimum of four wind sensors is to be operating with at least one based on a different operating principle to the others (ultrasonic vs. mechanical).

7.5 Fire and Flood Protection (EHS-F)

The **EHS-F** notation is supplement for **DPS-2** vessels with fire and flood protection implemented based on the fire risk level. This information is not necessary for **DPS-3** vessels, since **DPS-3** notation has higher requirements for fire protection.

7.5.1 Physical Separation

Physical separation is to be provided between redundancy groups, A-60 separation for machinery spaces of high fire risk and A-0 for other locations along the boundary of redundancy groups. The high fire risk area is the area defined by Class Rules including:

- Machinery spaces as defined by Class; and
- Spaces containing fuel treatment equipment and other highly flammable substances Watertight bulkhead separation is to be in place below the damage waterline for machinery spaces.

Power system. Each redundancy group is to have its own service tanks and it is to be located in individual compartments separated by A-60 partitions and be watertight if below the damage waterline. Each switchboard serving a different redundancy group is to be located in individual compartments separated by at least A-0 partitions and be watertight if below the damage waterline.

DP control station. Main and back-up (or additional) DP control stations are to have separate spaces, but A-0 partition is not required. The controllers, not located in DP control station, for the main DP control system and backup (or additional) DP control system are to be located in separate spaces with at least A-0 partitions. The redundant communication networks for the main DP control system are to be physically separated with A-0 partitions in general, A-60 in high-risk fire space and watertight if below the damage waterline.

Thrusters. Thrusters are to be located in separate watertight compartments with at least A-0 separation, provided fire risk escalation is low and active fire protection fitted in each compartment.

7.6 Summary of System Requirements for Enhanced System Notations

Enhanced system notations (**EHS** series) are complementary to the corresponding **DPS** series notations. The requirements for the **DPS** series notations given in Chap. 1, Table 7.1 are to be complied with. In addition, the listed requirements in Table 7.1 are also to be complied with for the Enhanced System notations (**EHS** series).

7.7 Closed Bus and Standby Redundancy

The DP system with enhanced electrical (**EHS-E**) and enhanced power and thruster notation (**EHS-P**) could have flexibility of closed bus operation and redundancy depending on standby start of the systems. The included in FMEA. The MTS publication, "*DP Operations Guidance*", can be used in the development of Operations Manual regarding closed bus operation.

7.7.1 Closed Bus Operation

The closed bus-tie configuration is to be capable of preventing a black-out under relevant fault conditions. The compensating provisions in the DP system's redundancy concept are to be such that the minimum number of required running generators

Table 7.1 Summary of DP system requirements for EHS notations

EHS-E	EHS-P	EHS-C	EHS-F
	Autonomous generator set	2 + 1 backup DP control computers and controllers[4]	**Generators and prime movers** Separate compartments, A60 for high fire risk area Watertight below damage waterline
Bus tie breaker redundantly configured between each bus segment	Bus tie breaker redundantly configured between each bus segment	Wind sensors 3 + 1 in back up control station	
Enhanced generator protection[1,2]	Enhanced generator protection[1,2]	Vessel heading sensors 3 + 1 in backup control station	**Power distribution system** A0 between redundant groups. watertight below damage waterline
Enhanced power management[1,3]	Enhanced power management[1,3]	MRU 3 + 1 in backup control station	**Thruster system** A0 between redundant groups Watertight below damage waterline
	Autonomous thruster set[1]	Position reference systems 3 + 1 in backup control station[5]	**Controller space** A0 between redundant groups

Notes (1) Controls and supervision of those functions are to be integrated in the control centre with the centralised control system and are to be communicated with DP controls

(2) Enhanced generator protection functions to be provide per *Enhanced system for* **DPS-2** *and* **DPS-3**

(3) Enhanced Power Management capability to be provided per *Enhanced system for* **DPS-2** *and* **DPS-3**

(4) For **DPS-2** vessels, manual position control system can be replaced with an additional DP control system as per *Enhanced system for* **DPS-2** *and* **DPS-3**. DP data logger is to be included as per the section entitled *DP control computers*

(5) Position reference systems are to be based on at least two different design principles. For deep-water conditions, GPS and Hydro Acoustics may be used

(i.e. two or more) are to be connected to two or more sections of the main bus. The minimum number of running generators are to be determined from the unit's DP capability. The worst-case failure of the configuration is not to result to a blackout. The total number of running generators is to be determined based on the power requirements, needed spinning reserve and the ability to phase back other loads. If bus frequency or voltage is outside of predefined limits the tie breakers are to open and the system is to run as a split system. The power system is to automatically subdivide for any failure effect that has the potential to defeat the redundancy design intent. Any DP system designed for operation with closed bus ties are also to be capable of being configured and operated with a defined performance and post failure capability in open bus ties configuration.

7.7.2 *Standby Start*

The standby start of the components or sub-systems of the DP system may be acceptable. The changeover operation is to be automatic and reliable provided that the position and heading of the vessel are within the specified limits and DP performance is not degrading. If no specific limit is specified, the maximum allowed changeover time is 45 s. The availability of standby start functions is to be monitored at the DP control station and communicated with the DP control system. In addition, the following factors are to be considered when using standby start as redundancy:

- Single fault does not cause total blackout including loss of entire compartment for **DPS-3**.
- A failure in one redundancy group is not to cause failure of any other redundancy groups.
- A failure in the system being changed over to is not to cause failure of any other redundancy groups.
- Changeover is not to cause failure of the redundancy group that is being connected to.
- One standby generator, at a minimum, is to be considered not available when needed.
- Redundant components or systems which are providing a standby function are not to share auxiliary systems or other services with the equipment they are providing standby redundancy for.

For a situation where thrust is to be immediately available (e.g., pipe laying) the redundancy is not to be based on standby start and changeover.

7.7.3 *FMEA and Testing*

Chapter 1 provides the requirements for FMEA analysis and FMEA proving trial for **DPS-2** and **DPS-3** notations. In addition to the requirements in Chap. 1, the enhanced features included in **EHS** series notations are to be analysed in FMEA and verified through testing. In addition to the failure modes provided in Chap. 1, for **EHS** series notations, following failure modes, where applicable, are also to be considered in FMEA analysis.

- Operation of protection systems (breakers, bus ties, etc.) related to short circuit.
- Severe voltage dips associated with short circuit faults in power plant configured as a common power system.
- Failure to excess and insufficient fuel.
- Over and under-excitation.
- Governor and AVR failure modes.
- Failure modes related to standby start and changeover.

- Power management failure on load sharing, malfunction, etc.
- Phase back thrust and large load.
- Blackout recovery.

The above required analyses are to be supported with extensive verification by testing. The test procedures are to be based on the simulation of failures and is to be carried out under as realistic conditions as practicable.

7.8 Documentation

The documentation required in Chap. 1 are to be expended to include the enhanced features where applicable, such as, but not limit to, following documents,

- DP Operations Manual **(I, OB)**.
- System description including a block diagram showing how the various components are functionally related **(R)**.
- Power generation and distribution **(R)**.
- Power analysis **(R)**.
- Description of Power management **(R)**.
- Thruster control system **(R)**.
- Auxiliary system for power generation and thrusters **(R)**.
- DP control system **(R)**.
- Failure modes and effects analysis (FMEA) **(R, OB)**.
- Test procedure for FMEA **(R)**.

In addition, the following additional documents are to be provided where applicable:

- Description of protection design philosophy and protection systems the redundancy concept of DP system depends on **(R)**.
- Analysis of effects of severe voltage transients on power system stability **(R)**.
- Short circuit analysis **(R)**.
- Simulation of severe over/under voltage and over/under frequency faults to prove the robustness of the power plant and its protection scheme for closed bus operation **(R)**.
- Protection coordination analysis **(R)**.
- Documentation of protection settings **(R)**.
- Description of automatic blackout recovery **(R)**.

Chapter 8
Station Keeping Performance

At the Vessel Owner's request, a station keeping performance notation (**SKP**), which s supplemental to the **DPS** series notations that may be assigned. The main objective of the station keeping performance notation is to provide for the consistent assessment of the station keeping performance reflecting the latest technology and to encourage robust redundancy design of DP systems. The station keeping performance (**SKP**) notation can be granted if the analysis has been carried out according to this Section and the results satisfy the requirements. Station keeping performance (**SKP**) may be assigned to each of the **DPS** series notations as supplemental information about the DP system. There are two levels of station keeping performance notations: **SKP** and **SKP (a,b,c,d,e,f)**. Notation **SKP** is for the station keeping assessment for given limiting environmental conditions and the assessment results meet the requirement of this book. Notation **SKP (a,b,c,d,e,f)** provides more information other than limiting environmental conditions in terms of probability that the vessel can remain on station for a given site environment.

8.1 Definition

8.1.1 SKP

SKP is an optional notation as supplemental information about the station keeping performance for specified limiting environmental conditions, such as wind speed and direction, wave height and frequency, current speed and direction. In general, the limiting environments are to be applied to intact (all thrusters running) and damaged (worst case failure) conditions unless the limiting environmental conditions for post failure cases are also specified. The directions for wind, wave and current can be different and are to be specified by the Vessel Owner. For example, a 1-year return

F. Karkori, *Dynamic Positioning Systems*, Springer Series on Naval Architecture, Marine Engineering, Shipbuilding and Shipping 21,
https://doi.org/10.1007/978-3-031-59173-0_8

environment is often used for MODUs as the limiting environment for DP operations. In that case **SKP** indicates that the station keeping performance of the vessel under the 1-year return environmental conditions has been verified through analysis, following procedures given in this Section.

8.1.2 SKP (a,b,c,d,e,f)

SKP (a,b,c,d,e,f) is an optional notation as supplemental information about the station keeping performance for a given environmental location. The analysis is to be carried out for many combinations of wind speeds, wave height and current speeds. The wind speed and wave height relationship is to be provided for the specified location. A co-linear condition for wind, wave and current is assumed.

a. The probability that the vessel can remain on station at the selected site **f** and current speed of **e** with all thrusters/rudders in normal operation conditions.
b. The probability that the vessel can remain on station with the failure of minimum effect single thruster/rudder at the selected site **f** and current speed of **e.**
c. The probability that the vessel can remain on station with the failure of maximum effect single thruster/rudder at the selected site **f** and current speed of **e.**
d. The probability that the vessel can remain on station at the selected site f and current speed of **e** with the worst case failure condition.
e. Current speed in knot (Vessel Owner-specified or typically 1.5 kt).
f. Environment location (Vessel Owner-specified or a representative typically North Sea location).

For example, **SKP (95, 95, 85, 75, 2, North Sea)** stands for

- 95% of time the vessel can remain on station with a current speed of 2-kt, North Sea wind and wave conditions, all thrusters operating,
- 95% of time the vessel can remain on station with a current speed of 2-kt, North Sea wind and wave conditions, with failure of minimum effect single thruster,
- 85% of time the vessel can remain on station with a current speed of 2-kt, North Sea wind and wave conditions, with failure of maximum effect single thruster, and
- 75% of time the vessel can remain on station with a current speed of 2-kt, North Sea wind and wave conditions, with worst case failure condition.

8.1.3 Normal Operation Condition

This is a designed condition for normal operation of a floating unit. The operating condition may be designed for one specific draft, or more than one operating drafts. If there are more operational conditions, minimum and maximum operational drafts need to be considered in the DP capability assessment.

8.1.4 Standby Condition

The standby condition is a condition when the intended operation is stopped voluntarily or involuntarily due to weather or other conditions. For an offshore drilling unit, the standby condition is a condition when risers are disconnected.

8.1.5 Traditional DP Capability Plot

In the traditional DP capability plot, the limiting environmental conditions are determined by varying wind speed for a given thruster configuration. Current speed is given and wave height is related to the wind speed.

8.1.6 Thrust Utilisation Plot

In the thrust utilisation plot, the thrust demand is calculated for a given limiting operating environmental condition. The wave height, wind speed and current are predefined. These types of plots allow the individual input of wind, current and wave data (in magnitude and direction) and display the power required for the thrusters over a 360° heading angle of the vessel.

8.2 Environmental Conditions

The environmental conditions are to be specified by the Owner of the DP vessel. The maximum design environments are generally to be determined by annual statistics.

8.2.1 Wind

Wind is treated as a constant in direction and speed. The wind speed is to be specified using one minute mean at height of 10 m above water surface. A value of 1.18 could be used to convert hourly mean wind speed to one minute mean speed.

8.2.2 Wave and Wave Spectrum

For **SKP** notation, significant wave heights and characteristic periods (frequencies) are to be provided. For notation **SKP (a,b,c,d,e,f)**, the relationship of wind speed and wave height and the probability of non-exceedance of each given wind speed are to be provided. The wave drift force is to be calculated for irregular sea states which can be described by a wave spectrum. A wave spectrum depends on geographical areas. The Owner of the DP vessel may specify the most appropriate wave spectrum determined from metocean data for the intended operating locations. If a site-specific wave spectrum is not available, the JONSWAP wave spectrum can be used for the North Sea and operating locations with limited fetch area. The JONSWAP wave spectrum can be expressed as:

$$S(\omega) = \frac{ag^2}{\omega^5} exp\left[-\beta\left(\omega_p/\omega\right)^4\right]\gamma^a$$

where

S wave energy density, in m^2/(rad/sec) (ft2/(rad/sec))
α 0.0081
g standard gravity, in m/s^2 (ft/sec2)
ω angular frequency of wave component, in rad/sec
β 5/4
ωp peak frequency, in rad/sec
γ peak enhancement factor
a $exp\left[\left(\omega - \omega_p\right)^2/2\omega^2{}_p\sigma^2\right]$
σ $0.07 if \omega \le \omega_p$
σ $0.09 if \omega > \omega_p$

For open seas, Bretschneider spectrum can be used, and the spectrum is given by:

$$S_\varsigma(\omega) = \frac{5\omega_p^4 H_S^2}{16\omega^5} \exp\left[-1.25\left(\omega_p/\omega\right)^4\right]$$

where

S_ς wave energy density, in m^2/(rad/sec) (ft^2/(rad/sec))
H_S significant wave height, in m (ft)
ω angular frequency of wave component, in rad/sec
ω_p peak frequency, in rad/sec.

8.2.3 Current

The current velocity is to include components due to tidal current, storm surge current, and wind driven current. The Owner of the DP vessel may specify the current speed determined from metocean data for the intended operating locations. For a MODU, the current profile needs to be provided for the calculation of current load on risers if deemed important, especially for those geographic areas where current force can be the governing design load.

8.2.4 Environments for SKP (a,b,c,d,e,f) Notation

A typical **SKP (a,b,c,d,e,f)** notation normally refers to the use of a set of representative North Sea wind and wave conditions refer to Table 8.1) and 1.5-knot current. For a **SKP (a,b,c,d,e,f)** notation for particular locations, for example the Gulf of Mexico, the site-specific information for current speed, non-exceedance probability of wind speed (wave height) and relationship between wind speed and wave height are to be provided. The site-specific environmental conditions are to be submitted for review.

Table 8.1 Relationship between wind and wave (North Sea)

Hs (m)	Tp (s)	Vw (m/s)	P non exc
0	0	0	0.0
1.28	5.3	2.5	14.3
1.78	6.26	5	35.7
2.44	7.32	7.5	56.9
3.21	8.41	10	73.3
4.09	9.49	12.5	85.2
5.07	10.56	15	92.5
6.12	11.61	17.5	96.6
7.26	12.64	20	98.4
8.47	13.65	22.5	99.3
9.75	14.65	25	99.7
11.09	15.62	27.5	99.9
12.5	16.58	30	99.9
13.97	17.53	32.5	100.0
15.49	18.46	35	100.0

Note Significant wave height, peak period and wind speed are based on Table 3 of "Specification for DP Capability Plot" by IMCA. The probabilities of non-exceedance are based on BMT Global wave statistics

8.2.5 Environments for SKP Notation

For the **SKP** notation, the limiting environmental conditions including wave height and period, wind speed and current speed, and their corresponding directions are to be provided by the Vessel Owner. The co-linear condition of wind, wave and current is to be assumed if direction information is not provided. For a MODU, the conditions for operation and standby need to be provided. In general, the 1-year environmental condition may be used for the operational condition.

8.3 Analysis Conditions

8.3.1 DP System Configurations

As a minimum, two conditions need to be considered:

- All thrusters in normal operating condition.
- The worst-case failure condition.
- The worst-case failure condition is to be determined from the basic design of redundancy concept.

8.3.2 Operating Conditions

Two operating conditions are to be considered as applicable:

(1) Normal operating condition. If a vessel is designed for operating at different loading conditions, namely different drafts, these are to be included in the station keeping analysis.
(2) Standby condition where applicable.

8.4 Environmental Load Calculation

Model test data on wind, wave and current load are to be used whenever possible. If however no model test results are available, the loads can be calculated according to this book.

8.4.1 *Wind and Current Force*

8.4.1.1 Non-ship Shape Unit

For a non-ship shape unit, wind and current force can be calculated according to the relevant Class Rules for building and classing mobile offshore units or the appropriate Class Rules for building and classing floating production installations, where applicable, or API RP 2SK.

8.4.1.2 Ship Shape Unit

For a ship shape unit, wind and current force can be calculated according to API RP 2SK or the OCIMF publication, "*Prediction of Wind and Current Load on VLCCs*".

8.4.1.3 Current on Drilling Risers

For a MODU, current force on drilling risers is to be included where deemed important.

8.4.2 *Wave Force*

Second order wave drift forces and yaw moment are to be considered in the station keeping capability analysis. They are to be calculated by using an appropriate hydrodynamic analysis computer program or extrapolated from model test results of a similar vessel.

8.5 Other External Loads

Where applicable, other external loads, such as pipeline tension, mooring line tension, heavy lift, may be considered for the DP capability assessment.

8.6 Available Thrust

8.6.1 Available Thrust for Thrusters

Manufacturer's test data of full scale or suitable model test for the thrust output of thrusters are to be used as the basis of **SKP** notation in general. The assessment of available thrust is typically indicated as thrust at full rated power. The availability of full power is a function of the prime mover characteristic. The power/torque/rpm characteristic for diesel driven thrusters is to be evaluated to determine the level of power available during DP operations (i.e., during zero inflow velocity). For thrusters with controllable pitch propellers as well as with hydraulically driven prime movers, full power is to be available at any inflow velocity. Electric drives typically have a certain RPM range in which full power is available. The thruster propeller is to be selected so that DP operations fall within this range. The selection of the propeller pitch should consider this range for optimum system efficiency. If no test results are available for the thrust output of thrusters, Annex 1 of this book provides the guidelines for determining available thrust, which deals with typical thrusters and installation scenarios for DP vessels. Those guidelines may be used for preliminary studies.

8.6.2 Thruster-Thruster Interaction

When one thruster operates downstream of another, the available thrust of the thruster is reduced due to thrusters interaction. The effect of the interaction depends on the following

- Distance between the thrusters.
- Azimuth of the thrusters.
- Diameter of the thruster propeller.
- Thruster load.
- Thruster design/configuration (i.e., degree of tilt of the propeller and/or nozzle axis).

This interaction effect is to be included in the station keeping performance assessment. The results from full scale or suitable model test for the thrust-thrust interaction effect are to be used whenever possible. If no such results are available, Annex 1 of this book can be used as guidelines for the assessment of the interaction effect on the available thrust.

8.6.3 Thruster-Hull Interaction

The interaction between thrusters and the hull also reduces available thrust of the thrusters. The interaction includes following factors, and they are to be included in the station keeping performance analysis.

- *Friction*. The flow of the slipstream along the hull will result in the thrust degradation due the friction of the hull. The degradation is related to the length and breadth of the flow along the hull.
- *Coanda effect*. When a thruster is oriented in a transverse direction, the output thrust of the thruster is affected by a so called Coanda effect. The reduction of the thrust is related to the bilge radius and the length of the flow underneath the hull.
- *Pontoon blockage*. The blockage of the slipstream due to presence of the pontoon, such as when a slipstream is orientated toward the pontoon, will affect the thrust output of the thruster. The effect is related to the distance between the pontoons and the azimuth of the thruster.
- *Tilted thruster/nozzle*. A tilted thruster/nozzle can reduce above mentioned thruster-hull interactions and hence to improve the output of the thrust of the thrusters. This improvement can be considered in the station keeping performance analysis.

The full scale or model test results for the effects of thruster-hull interactions mentioned above and the tilted thruster/nozzle are to be used whenever possible. If such results are not available, Annex 1 of this book can be used as guidelines for the assessment of the interaction effect on the available thrust.

8.6.4 Thruster-Current Interaction

Current inflow could reduce the thrust output of the thrusters and the thrust reduction is to be included in the station keeping performance analysis. Manufacturer's test data of full scale or suitable model test for the current effect is to be used whenever possible. It such data is not available, Annex 1 of this book can be used as guidelines for the assessment of the current effect.

8.6.5 Tunnel Thruster

The available thrust generated by a transverse tunnel thruster is highly affected by the tunnel installation, location, and the geometry and configuration of the hull. The assessment of the tunnel thruster performance is to be based on axial flow pump hydrodynamics, not on marine propeller theories. Manufacturer's test data of full scale or suitable model test for the efficiency of the tunnel thruster is to be used whenever possible. It such data is not available, Annex 1 of this book can be used as guidelines for the assessment of the efficiency of the tunnel thrusters.

8.7 Rudder Forces

For vessels with conventional propulsion arrangements, the propeller jet and its interaction with the rudder can be used to develop transverse forces in the stern of the vessel in lieu of thrusters during dynamic positioning operations. When this approach is used, the following considerations are to be addressed:

- Control over propeller thrust from zero to full power is required. This requires the installation of a controllable pitch propeller for diesel engine driven systems.
- Only twin-screw propulsion systems are to be considered for generating transverse rudder forces. With most rudders (exception: some steerable nozzle designs), the generation of transverse force simultaneously generates an unwanted force in the ahead direction. This parasitic thrust is to be compensated by thrust delivered in the astern direction by operating the second propeller (or by other azimuth thrusters, if available).
- The use of conventional rudders for DP yields to very limited side forces. High-performance rudders or steerable nozzles are preferred for DP applications.

Manufacturer's test data of full scale or suitable model test for the rudder forces are to be used whenever possible. If such data is not available, Annex 2 of this book can be used as guidelines for the assessment of the rudder forces.

8.8 SKP Calculations

The calculations for **SKP** are based on the balance of horizontal forces and yaw moment while the vessel is maintaining both position and heading. Those forces and moment include all forces and yaw moment resulted from thrusters, wind, waves, current and other external forces where applicable. Where the vessel has been subject to alteration or addition, which may affect the station keeping characteristics of the DP system, the **SKP** is to be recalculated.

8.8.1 *SKP Notation*

For the **SKP** notation, the station keeping performance can be carried out by using quasi-static analysis approach or time-domain simulation approach that includes the dynamic component of wave load, controller and thruster response time. The station keeping performance assessment is to include the following:

- Limiting environmental conditions specified by the Vessel Owner.
- Co-linear condition of wind, wave and current.
- Different directions for wind, wave and current specified by Vessel Owner.
- Analysis for 10-degree interval of headings between 0 to 360 degrees if not specified by the Vessel Owner.
- Analysis for the heading ranges specified by the Vessel Owner.

8.8.2 *SKP (a,b,c,d,e,f) Notation*

For the **SKP (a,b,c,d,e,f)** notation, the requirements for **SKP** is to be met if applicable. In addition, following conditions are to be considered.

- Relationship of wave height, period and wind speed to be provided.
- **SKP** numbers of **(a, b, c, d)** are calculated considering all heading station keeping capabilities.
- Capability calculation for 10-degree interval of headings between 0 to 360 degrees.
- The **SKP** numbers are the weighted sum of all heading results.
- If no heading probabilities are specified, equal probability for all headings can be used.

8.9 Documentation

Where the DP system is to be supplemented with a station keeping capability notation, the following information is to be submitted for assignment of a **SKP**.

- General arrangement and lines plan.
- Thruster arrangement.
- Thruster power and thrust.
- Thruster interactions.
- Capability plots.
- Documentation on the environmental conditions long term distribution (any area for intended service).
- Relationship between wind speed and wave height if applicable.
- Vessel Owner specified limiting environmental conditions.

Chapter 9
Specific Vessel Types

While station keeping using DP is a common function of all DP vessels, other functions driven by specific activities that a vessel will undertake are also to be considered. Additional criteria, specific to several vessel types, are given below. Mobile offshore drilling units, project or construction vessels, and logistics vessels are addressed, respectively, in later sections of this chapter.

9.1 Mobile Offshore Drilling Units

9.1.1 Basic Design Criteria for DP System

9.1.1.1 Fault Tolerance

For DP MODUs, **DPS-2** equipment class and above are to be considered excluding self-elevating units that may use the DP system and its thrusters to maintain position only when installing the unit on-site (e.g. setting jack-up legs).

9.1.1.2 Limiting Environmental Conditions

Although the environmental conditions are specified by the Vessel Owner of the DP MODU, in general, one-year environmental condition for the area of operation is to be considered for DP MODU operating conditions. For position standby conditions, the design environmental condition is normally higher than that for operating conditions.

F. Karkori, *Dynamic Positioning Systems*, Springer Series on Naval Architecture, Marine Engineering, Shipbuilding and Shipping 21,
https://doi.org/10.1007/978-3-031-59173-0_9

9.1.1.3 Station Keeping Performance

Station keeping performance is to be assessed for the limiting environmental conditions for operating and for position standby conditions. Station keeping performance assessment is also to be performed for intact and post worst-case failure conditions. The current load on risers is to be considered, if deemed important.

9.1.2 Position Reference Systems and Sensors

The selection and installation of position reference systems and sensors are to take into account the suitability and reliability of the systems. Satellite and acoustics are two types of position reference systems that are suitable for deep water operation. The update rate of acoustics and the impact of possible obstructions for the satellite systems are to be considered when deciding the suitability of the position reference system. Enhanced DP control system (**EHS-C**), includes additional requirements for position reference systems with enhanced availability and reliability, refer to Chap. 7, *Enhanced System*, for more information.

9.1.3 Effect of Drilling Loads

Power consumption of drilling operations and power sharing between the DP system and drilling operations are to be analysed and submitted to Class for review.

9.1.4 Vessel Operating Condition

In addition to the normal drilling operational condition, positioning standby condition is also to be considered in the DP system design in terms of available power to the DP system and the station keeping capability.

9.1.5 Emergency Shutdown System

The general emergency shutdown philosophy for the vessel and the effect of the emergency shutdown system on the redundancy of the DP system is to be carefully considered. Risks associated with technical faults and inadvertent spurious operation after inadvertent operations of the emergency shutdown system are to be considered. Each vessel is to develop a detailed plan for recovery and restoration of operation after operation of each level of ESD. ESD Stations that can enable a total unit shutdown are not to be located in locations which are unmanned under normal operations except in the backup DP control station, if provided. Where ESD stations are provided at the lifeboat stations or other unmanned locations, the total unit ESD (complete shutdown) is to be protected from unauthorised personnel or not available at these unmanned locations.

9.1.5.1 ESD Operation Manual

- The ESD operation manual describes the ESD system and unit specific operational guidelines.
- ESD Operations are to consider the potential risks and risk areas as well as the appropriate responses for each risk.
- The ESD Operation Manual is to define the ESD levels and provide a list of equipment or areas that are affected by the different ESD levels. Also, the manual is to indicate which ESD levels are available at each ESD station. Further, the manual is to provide instructions on to reset the affected systems after each ESD.
- The descriptions and instructions in the ESD Operation Manual are not to conflict with the WCFDI identified in the DP FMEA.
- The ESD Operation Manual is to provide guidance describing the typical scenarios that the ESD levels are to be used and who has access to use them.
- The ESD Operation Manual is to be included or referenced in the unit's operating manual.

9.1.5.2 Gas Detection/ESD System Cause and Effect Chart

Where shutdown groups are initiated automatically upon gas detection, a Gas detection/ESD system cause and effect chart shall relate gas detection sensors to ESD shutdown groups of equipment and areas on the unit.

9.1.6 Emergency Disconnect System

Time for safe disconnection of lower marine riser package (LMRP) is to be considered in the DP system design for the post failure station keeping capability. The operational impact of working in suitable water depth is to be considered in terms of required safe disconnecting time and the feasibility of the equipment and procedures is to be studied.

9.1.7 Maintenance Plan

DP MODUs is to have a structured planned maintenance system that specifically addresses maintenance of the DP system, equipment and support systems. Planned maintenance is to address all equipment that has an impact on the vessel's station keeping capabilities. This is to include indirect components such as generator circuit breakers, bus tie breakers, etc. Maintenance is to include regular cleaning, calibration, and testing of equipment as outlined in manufacturers' recommendations and industry guidelines. Records of planned and unplanned maintenance are to be kept in an auditable format, either hard copy or appropriate electronic format. These records are to include vendor service records as well as maintenance performed by vessel personnel. These records are to be kept onboard for the period specified by the Vessel Owner, and they are to be available for review by the Class approved Surveyor during periodic surveys of the vessel to maintain the DPS related notation.

9.1.8 FMEA

In addition to the DP system, the DP FMEA for a DP MODU is to take account of non-DP systems that have interfaces with and potential impact on DP and station keeping, such as:

- Drilling system.
- Emergency shutdown system.
- Safe disconnection of risers.
- Lower marine riser package (LMRP).

9.1.9 DP Alert System

A system of visual and audible alarms, supplied by a UPS and with an indication of power being available, is to be provided in each DP control station, on the navigation bridge, at the engine control position, at drill floor, and at drilling supervisor's office. The status of the DP system in operation is to be indicated.

9.1.10 DP Operations Manual

In addition to the requirements outlined in the section *DP operations manual* in Chap. 1, *Dynamic Positioning System Design*, the DP operations manual for a MODU is to address operating issues relating to equipment and systems that are interfaced to the DP system. This includes ESD systems installed on DP MODUs and emergency disconnect system. Detailed procedures are to be developed for operating ESD and EDS. Procedures are to highlight criteria for initiation and consequences. The MODU's DP operating manual is to clearly identify where this information can be found if not included in the manual. The MTS publication "*DP Operations Guidance, Part 2, Appendix 1 (DP MODU)*" can be followed for the development of the DP Operations Manual.

9.1.11 Documentation

In addition to the documentation requirements outlined in section *Alarms, monitoring and instrumentation* of Chap. 1, *Dynamic Positioning System Design*, the following documents are to be submitted for review:

- ESD operation manual.
- Gas detection/ESD system cause and effect chart.

9.2 Project or Construction Vessels

9.2.1 Basic Design Criteria for DP System

9.2.1.1 DP Control System

The DP control system is to be equipped with suitable DP modes for the activities carried out by the project/construction vessels. When carrying out industrial missions, the availability of the suitable DP mode is to be verified. The following selected DP control modes are relevant to specific DP activities.

- *Target follow:* Enables the DP vessel to follow a moving target.
- *Heavy lift:* Takes account of the effects of the load transfer on the mass of the vessel and the additional lateral force.
- *External force compensation:* Where the measured external force acting on the vessel, which is separate from the environment, is included in the DP calculation and treated as a force feed forward. This mode is used to account for pipe tensions in a pipe laying vessel and hawser tension in a shuttle tanker.
- *Fire monitor compensation:* Used to compensate for the varying forces exerted on a vessel from discharge of a fire monitor.
- *Weathervane:* Enables the DP vessel to rotate with the wind, current and waves around a fixed or moving point called the terminal point. Neither the heading nor the position of the DP vessel is fixed. The heading of the vessel is controlled

to point towards the terminal point. The position of the vessel is controlled to follow a circle, called the setpoint circle, around the terminal point. This mode is appropriate for connected shuttle tanker/FPSO operations.

9.2.1.2 Station Keeping Performance

External forces and weathervane are to be included in the station keeping performance assessment where applicable.

9.2.2 Operation Configuration

In general, the DP project/construction vessels are to operate in open bus. If a closed bus tie is one of the operation configurations, risk assessment is to be performed. Validation testing is to be carried out to demonstrate that the common point introduced by the closing of the bus ties does not provide a fault propagation pathway and equipment related to DP Station keeping (including auxiliary systems and supporting systems) demonstrate voltage dip fault ride through capability.

9.2.3 Position Reference Systems and Sensors

The selection and installation of position reference systems and sensors are to take into account the activities carried out by the DP vessels and the obstructions from other units. For project/construction DP vessels, redundancy of the relative reference systems is to be considered when carrying out activities close to other structures and the loss of the relative reference system is to be included in the FMEA. Satellite and acoustics are two types of position reference systems that are suitable for deep water operation. The update rate of acoustics and the impact of possible obstructions on the satellite systems are to be considered when deciding the suitability of the position reference system.

9.2.4 Maintenance Plan

For project/construction vessels where preventive maintenance is to be carried out while undertaking project activities, due consideration is to be given to methods that will permit the vessel to continue the operation in the identified operation mode. The details are to be included in the Operations manual.

9.2.5 FMEA

In addition to the DP system, the Overall Vessel FMEA for a DP project/construction vessel is to take account of non-DP and auxiliary systems, if applicable, that have interfaces with and potential impact on DP and station keeping.

9.2.6 DP Alert System

For diving support vessels, a system of visual and audible alarms, supplied by a UPS and with an indication of power being available, is to be provided in each DP control station, on the navigation bridge, at the engine control position, in the dive control, saturation control and air diving control areas, in the working area, and, where applicable, the ROV or submersible control position. The status of the DP system in operation is to be indicated.

9.2.7 DP Operations Manual

In addition to the DP system, the DP operations manual for a project/construction vessel is to address operating issues relating to equipment and systems that are interfaced to the DP system, if applicable. An *Activity Specific Operation Guideline* (ASOG) is to be developed for each activity undertaken by the vessel and be referenced in the DP operations manual.

The MTS publication, "*DP Operations Guidance, Part 2, Appendix 2 (Project and Construction Vessels)*", can be followed for the development of the DP operations manual.

9.3 Logistics Vessels

9.3.1 Operation Mode

The DP logistics vessel mainly provides logistics support for the project. The vessel, in general, works in DP mode in a 500 m (1,640 feet) proximity operation zone. Generally, closed bus tie operation is to be avoided for logistics vessels. If closed bus tie is one of the operation configurations, risk assessment is to be performed. Validation testing is to be carried out to demonstrate that the common point introduced by the closing of the bus ties does not provide a fault propagation pathway and equipment related to DP Station keeping (including auxiliary systems and supporting systems) demonstrate voltage dip fault ride through capability.

9.3.2 Position Reference Systems and Sensors

The selection and installation of position reference systems and sensors are to take into account obstructions from other units. For DP logistics vessels, loss of the relative reference system is to be included in the FMEA for the activities to be carried out. The redundancy of the relative reference systems is to be considered where necessary and is to be consistent with the redundant design of the DP system.

9.3.3 DP Operations Manual

The MTS publication, "*DP Operations Guidance, Part 2, Appendix 3 (Logistic vessels)*", can be followed for the development of the DP operations manual.

9.4 Articulated Tug-Barge Combinations

The requirements of this section apply to the application of dynamic positioning notations as outlined in this book for articulated tug-barge (ATB) combinations that comply with the Class Rules for building and classing steel barges (as relevant). The requirements of this book are to be applied except as modified herein. The barge and associated towing vessel are considered dependent upon each other to satisfy the dynamic positioning requirements when the two vessels operate as a combined unit in DPS mode. To apply the exemplar DPS notations as outlined in this book, the towing vessel is to be ABS classed and assigned the Towing Vessel ATB notation and the barge is to be ABS classed and assigned the ATB notation as per the relevant section of the Class Rules for building and classing steel barges. Accordingly, example notations for both the barge and associated towing vessel:

✠A1 Barge, ATB, DPS-0

✠A1 Towing Vessel ATB, DPS-0

Dynamic positioning systems built and tested in compliance with the requirements outlined throughout this book and accordingly relevant Rules may be assigned with different classification notations depending on the degree of redundancy built into the system as defined in Chap. 1 of this book. These notations are not a requirement for classification of the ATB and are to be assigned only on the specific request of the Vessel Owner. Where dynamic positioning is requested, both the barge and the associated towing vessel as part of the ATB combination are to receive the same selected **DPS** notation. In addition to the notations, the following comments will be maintained in the Class Record for both the barge and associated towing vessel in compliance with this book:

- Towing Vessel ATB:
  "Classed to operate in **DPS** mode as an ATB with the associated barge(s) (Vessel Name(s) and Class Number(s))".
- Barge ATB:
  "Classed to operate in **DPS** mode as an ATB with the associated towing vessel(s) (Vessel Name(s) and Class Number(s))".

In cases where either the towing vessel or barge request a **DPS** notation as an independent vessel in addition to an ATB combination, documentation and testing is to be conducted for both modes of DP operations. The proposed separate DP modes of operation are to be outlined in the DP operations manual. The above record comments are to reflect the separate independent and dependent ATB combinations as applicable. Where changes are made to the ATB combination for dynamic positioning operations, the owner is to notify Class. Multiple vessel combinations may be permitted with each combination configuration approved and recorded.

9.4.1 Documentation

The following documentation is to be submitted in addition to that required in Chap. 1 of this book:

- General arrangements of combined ATB unit (**R**, **OB**).
- DP operation manual (**I**, **OB**).
- ESD operation manual (**I**, **OB**).
- Maintenance plan (**I**, **OB**).
- Tug/Barge heading deviation analysis (**I**).
- Cable installation and arrangement details between barge(s) and associated towing vessels (**R**).

9.4.2 Certification

The components and subsystems associated with the dynamic positioning system as fitted on both the towing vessel and barge are to be certified or type approved for suitability in the marine environment in accordance with the Class Rules for building and classing marine vessels.

9.4.3 *Alternatives*

Equipment, components, and systems for which there are specific requirements in this book, or its associated references, may incorporate alternative arrangements or comply with the requirements of alternative recognised standards, in lieu of the requirements in this book. This is subject to such alternative arrangements or standards being determined by Class as being not less effective than the overall safety and strength requirements of this book or associated references. Where applicable, requirements may be imposed by Class in addition to those contained in the alternative arrangements or standards so that the intent of this book are met. In all cases, the equipment, component or system is subject to design review, survey during construction, tests and trials, as applicable, by Class for purposes of verification of its compliance with the alternative arrangements or standards.

9.4.4 *Tug Barge Arrangement*

9.4.4.1 Connection Between the Towing Vessel and Barge

Access

Readily available access between the barge and associated towing vessel is to be provided.

Cables between tug and barge

Details of the hard-wired connections made for the supply of electrical power, communications, control and monitoring between the barge and associated towing vessel are to be submitted.

Power connection box and cable

Where power is to be provided from the towing vessel to the barge or from the barge to the towing vessel, the requirements of the Class Rules for building and classing marine vessels.

Cables and connections

Cable and associated fittings between the barge and associated towing vessel are to meet the requirements of 4-8-3/9 of the Rules for Building and Classing Marine Vessels.

Installation

Cable installations between the barge and associated towing vessel is to meet the cable installation requirements of the Class Rules for building and classing marine vessels. Cables are to be installed and supported so as to avoid chafing and undue stress in the cable. Cable supports and associated accessories are to be robust and are to be of materials that are corrosion-resistant or suitably treated to resist corrosion and arranged so that flexure of the cable from the movements between the towing vessel and the barge is kept to a minimum. Cable installation and arrangement details between the barge and associated towing vessel are to be submitted.

9.4.4.2 DP Control System

The main DP control station as addressed by Chap. 4 of this book is to be provided on the associated towing vessel navigation bridge. Alarm, monitoring and instrumentation for systems fitted on the barge are to be provided on board the barge in a centralised control room and provided at the main DP control station on board the associated towing vessel. Instrumentation and controllers associated with the barge machinery are also to be provided at the main DP control station, as required by this book (refer to Chap. 1, *Dynamic positioning system design,* Table 2, *Summary of minimum instrumentation at the DP control station*) and in accordance with Table 9.1 of this chapter as applicable. Where power is supplied to the barge from the associated towing vessel for DP operations, the alarms, displays, and controls as reflected in Table 9.1 may be reduced. Items A1-A6, B1-B4, and D1-D2 are to be maintained at the main DP control station, as applicable.

9.4.4.3 Automation

The control, alarms, and monitoring associated with machinery fitted on the barge and associated towing vessel are to be accordance with the applicable Class Rules for building and classing marine vessels, as applicable. Separate from the DPS FMEA (**DPS-2/DPS-3**), an overall ATB combination FMEA is to be conducted where DPS control integration may have an impact on essential services and safety functions on either the tug or barge. An overall ATB combination FMEA is to be carried out as per the relevant sections of the Class Rules for building and classing marine vessels. The FMEA is to demonstrate that the integrated system will ‘fail-safe’, and that essential services in operation will not be lost or degraded.

9.4.4.4 Position Reference Systems

In addition to position reference systems and associated sensors required in the section *Position reference system and environment sensor,* in Chap. 4, *Control system,* deviation of heading between the barge and associated towing vessel is to be considered. Analysis of heading deviation that may exist between the barge and associated towing vessel and the effect on the dynamic position capabilities is to be submitted.

Table 9.1 Instrumentation and controllers at the main DP control station

System	Monitored/controlled parameter		A	D	C	Notes
System monitoring	A1	System power supply main and secondary feeders: failure, status, and transfer	x	x		Applicable to main and secondary power sources
	A2	Individual power supply to control, monitoring, and safety	x	x		Alarm may be common
	A3	Alarm system—disconnected		x		
	A4	Integrated computer-based system: data highway abnormal conditions	x			Alarm is to be activated before critical data overload
	A5	Integrated computer-based system: duplicated data link—failure of one link	x			
	A6	Automatic control systems	x	x		
Others	B1	Control station transfer		x	x	Display: to indicate the station in control Control: to provide transfer switch acknowledgment switch
	B2	Air conditioning system—fails	x			If necessary for equipment environment control
	B3	Machinery space—fire detected	x	x		
	B4	Machinery space—bilge level high	x			
Electric power generating plant	C1	Starting, paralleling and putting generator online			x	
	C2	Generator running		x		
	C3	Voltage—high and low	x	x		
	C4	Current—high	x	x		
	C5	Frequency—high and low	x	x		
	C6	Failure of online generator	x			

(continued)

Table 9.1 (continued)

System		Monitored/controlled parameter		A	D	C	Notes
		C7	Generator engine auxiliaries start/stop		x	x	Automatic start/stop, if fitted, is to be alarmed
		C8	Bearing lube oil inlet pressure—low	x	x		Automatic shutdown prime mover
		C9	Generator cooling inlet pump or fan motor—fails	x			
		C10	Generator cooling medium temp—high	x	x		
Power distribution system		D1	Status of automatically controlled circuit breakers		x		
		D2	UPS fault	x			
Fuel oil system		E1	Service tank level—low	x			
Diesel engine	Lubricating oil	F1	Bearing oil inlet pressure—low	x	x		Automatic shutdown with alarm at low-low
		F2	Bearing inlet oil temperature—high	x	x		
		F3	Common rail servo oil pressure—low	x			
	Cooling medium	F4	Pressure or flow—low	x	x		
		F5	Temperature at outlet—high	x	x		
		F6	Expansion tank level—low	x			
	Fuel oil	F7	Fuel oil leakage from injection pipe	x			
		F8	Common rail fuel oil pressure—low	x			
	Starting medium	F9	Energy level—low	x	x		For application requiring automatic generator paralleling
	Speed	F10	Overspeed	x			Auto shutdown

A = alarm
D = display
C = controller/actuator
X = applies

9.4.4.5 Emergency Shutdown System

The general emergency shutdown philosophy for the vessel and the effect of the emergency shutdown system on the redundancy of the DP system is to be carefully considered. Risks associated with technical faults and inadvertent operations of the emergency shutdown system are to be considered. Each vessel is to develop a detailed plan for recovery and restoration of operation after operation of each level of ESD. ESD stations that can enable a total vessel shutdown are not to be located in locations which are unmanned under normal operations except in the backup DP control station, if provided. Where ESD stations are provided at unmanned locations, the total unit ESD (complete shutdown) is to be protected from unauthorised personnel or not be available at these unmanned locations.

ESD operations manual

(1) The ESD operation manual describes the ESD system and unit specific operational guidelines.
(2) The ESD operation manual is to define the ESD levels and provide a list of equipment or areas that are affected by the different ESD levels. Also, the manual is to indicate which ESD levels are available at each ESD station. Further, the manual is to provide instructions on how to reset the affected systems after each ESD.
(3) The descriptions and instructions in the ESD operation manual are not to conflict with the WCFDI identified in the DP FMEA.
(4) The ESD operation manual is to be included or referenced in the DP and cargo operating manual, as applicable.
(5) Where shutdown groups are initiated automatically (e.g. automatic shutdowns related to cargo operations), a ESD system cause and effect chart shall relate ESD shutdown groups of equipment and areas on both the barge(s) and associated towing vessel.

9.4.4.6 DP Operations Manual

In addition to the requirements outlined in the section *DP operations manual*, of Chap. 1, *Dynamic positioning system design*, the DP operations manual for an ATB is to address operating issues relating to equipment and systems that are interfaced to the DP system. The articulated coupling systems may allow for vertical movement within the notch without completely disconnecting the two vessels during cargo transfer operations and station keeping operations. Consideration may also be given to operational methods and ballasting procedures proposed as an alternative for limiting the vertical loads imposed by the connection. These procedures are to be included in the DP operations manual and maintained on both the towing vessel and barge. In association with the section *Load line stability*, where the barge is issued an unmanned service load line, the DP operations manual is to reflect the barge is to operate with a freeboard not less than the calculated freeboard as a manned vessel. The DP operations manual is to be kept onboard both the tug and barge, and they are to be available for review by the Surveyor during periodic surveys of the vessel to maintain the associated **DPS** notation.

9.4.4.7 Maintenance Plan

An ATB is to have a maintenance plan that specifically addresses maintenance of the DP system, equipment and support systems on both the tug and barge. Planned maintenance is to address all equipment that has an impact on the vessel's station keeping capabilities. This is to include indirect components such as generator circuit breakers, bus tie circuit breakers, and other power distribution and control equipment associated with the DPS. Maintenance is to include regular cleaning, calibration, and testing of equipment as outlined in manufacturers' recommendations and industry guidelines. Records of planned and unplanned maintenance are to be kept in an auditable format, either hard copy or appropriate electronic format. These records are to include vendor service records as well as maintenance performed by vessel personnel. These records are to be kept onboard both the tug and barge, and they are to be available for review by the Class approved Surveyor during periodic surveys of the vessel to maintain the associated **DPS** notation.

9.4.5 Barge

9.4.5.1 Bilge Level Monitoring

Where a machinery space is located below main deck, means to detect excessive rise of bilge water in the bilges or bilge wells is to be provided and alarmed (refer to Table 9.1, *Instrumentation and controllers at the main DP control station*). The arrangements including the number of sensors and locations are to be such that accumulation of bilge water may be detected at the various angles of barge's heel and trim.

9.4.5.2 Fire Detection System

A fire detection system is to be provided for machinery spaces on the barge. Indication of the fire detection is to be provided at DP control stations. The fire detection system is to be in accordance with the Class Rules for building and classing marine vessels.

9.4.5.3 Fire Protection

The fixed fire extinguishing fire protection outlined in the section on *Fire protection* in Chap. 1, *Dynamic positioning system design*, is to protect internal combustion machinery spaces associated with DPS operations.

9.4.5.4 Lighting

Enclosed spaces associated with DP operations are to be provided with lighting and provided with backup power supply for a period of at least 90 min.

9.4.5.5 Passage of Crew

(1) Bulwarks and/or guardrails complying with the Class Rules for building and classing steel barges, as applicable, are to be provided.
(2) Secondary crew protection at or near the centreline in accordance with the *International Load Line Convention 1966, Regulation 25–1.*

Special consideration may be given to Flag Administration alternative requirements.

9.4.5.6 Load Line and Stability

Further to the Class Rules for building and classing steel barges, the barge while conducting DP operations, is to operate with a freeboard not less than the calculated freeboard as a manned vessel in accordance with the *International Convention on Load Lines, 1966* as amended. The barge is to be fitted with the features as required by the *International Convention on Load Lines, 1966* as amended, for a manned vessel. All intact and damage (where applicable) stability requirements are to be taken into account when assigning the freeboard. Special consideration may be given to Flag Administration alternative requirements. The level of manning required for the tug and barge falls outside the scope of Class and is to be determined by the Flag Administration.

9.4.5.7 Lifesaving Appliances and Equipment

Provide the following measures for the maximum number of personnel to be on the barge during DP operations:

(1) Survival craft.
(2) Personal lifesaving appliances.

Special consideration may be given to Flag Administration alternative requirements. The level of manning required for the barge and associated towing vessel are outside the scope of Class and is to be determined by the Flag Administration. Exemptions granted by or instructions received from the Flag Administration based on the operation distance from shore and number of persons on board and other emergency procedures will be considered by Class.

9.4.6 Surveys

The provisions in this section are requirements for obtaining and maintenance of classification of **DPS** notations. These requirements are in addition to the provisions noted in Chap. 7, *DP system initial test*. Both the barge and associated towing vessel are to be connected to demonstrate compliance. In cases where either the towing vessel or barge request a **DPS** notation as an independent vessel in addition to an ATB combination, the independent vessel is to demonstrate compliance in addition to demonstrating compliance an ATB combination.

9.4.6.1 Surveys During Construction

Surveys are to be carried out on both the barge and associated towing vessel during construction, installation, and testing of the asset at the builder's yard/facility, including required onboard testing and trials. DP system initial tests are to be conducted as per Chap. 7, *DP system initial test.*

9.4.6.2 Surveys After Construction

Surveys after construction for dynamic positioning are to be in accordance with the Class Rules for survey after construction. Both the barge and associated towing vessel are to be connected to demonstrate compliance. The barge and associated towing vessel is to be operated to demonstrate that the automatic and remote control systems have been maintained properly and are in good working order. The operational testing is to be carried out to the Class approved Surveyor's satisfaction. Records of planned and unplanned maintenance are to be kept in an auditable format, either hard copy or appropriate electronic format. These records are to include vendor service records as well as maintenance performed by vessel personnel. These records are to be kept onboard for the period specified by the Vessel Owner and are to be available for review by the Class approved Surveyor during periodic surveys of the vessel to maintain the **DPS** related notation.

9.4.6.3 Documentation

The following documents are to be confirmed onboard the barge and associated towing vessel as applicable:

- General arrangements of combined ATB unit.
- DP operation manual.
- ESD operation manual.
- Maintenance plan.

Chapter 10
DP Operations

This chapter provides DP operations guidance for the safe and effective conduct and management of DP operations. The guidance is based on the DP Operation Guidance by the Marine Technology Society's (MTS) DP Technical Committee. The DP Operation Guidance includes two parts:

- Part 1 contains guidance on the management of DP systems for the identified DP themes.
- Part 2 contains DP operational guidance in detail in three appendices:
  - Appendix 1-DP MODUs.
  - Appendix 2-DP Project/construction vessels.
  - Appendix 3-DP Logistics vessels.

Part 1 of the *DP Operation Guidance* addresses the DP themes that, if managed effectively, will contribute to safe and efficient DP operations. Fourteen important DP operation themes have been identified in Part 1. Part 2 provides operational guidance in detail on the themes identified in Part 1. This chapter presents the fourteen identified DP themes and brief discussion on the themes for the safe and efficient DP operations. Users are recommended to peruse *DP Operation Guidance* for details. The *DP Operation Guidance* provides recommended DP documentation that should be kept on board and, where feasible, at the shore-based centres of technical management. The list of the recommend documents is included at the end of the chapter.

F. Karkori, *Dynamic Positioning Systems*, Springer Series on Naval Architecture, Marine Engineering, Shipbuilding and Shipping 21,
https://doi.org/10.1007/978-3-031-59173-0_10

10.1 Themes

10.1.1 DP Class

DP systems have been grouped into three equipment classes in IMO MSC.1/Circ.1580 and IMO MSC Circ.645, namely DP class 1, DP class 2 and DP class 3, based on the level of redundancy of the components and systems. Selection of DP equipment class is mainly driven by the intended operations of the DP vessels. For most operations, DP class 2 notations are recommended. For some operations, such as ROV support, seismic and survey in open water condition, logistics operations, vessels of lower DP equipment class notations may be used. The fundamental principal of selecting DP class should be based on a proper risk assessment of the operations.

10.1.2 DP FMEA

The DP vessel's DP FMEA is the most important technical document and is a requirement to obtain DP class 2 and 3 notations. The development of the FMEA should be carried out according to the rules/Guides of the vessel's Classification Society. The FMEA is also to comply with the IMO Guidelines and achieve the standards of detail and analysis of industry guidance, such as related publications by IMCA and OCIMF. Specific attention is drawn to the *Dynamic Positioning Failure Mode Effects Analysis Assurance Framework Risk Based Guidance* (First Edition 2020) published by OCIMF. Key DP personnel, including the vessel Master, DPOs, Engineers and Electricians should have a detailed knowledge of the DP FMEA and should use the information provided to be fully informed about the capabilities and limitations of the vessel's DP system. The results from a DP FMEA, in particular issues related to the vessel's worst-case failure and significant single point failures, should be used in the formulation of operational, emergency response and planning decisions.

10.1.3 DP Capability Plots

DP Capability Plots should be calculated for the vessel according to the Rules/Guides of the vessel's Classification Society and recognised industry guidelines. The plots should be provided in a format that is intuitive to the user on board. DP Capability Plots may have to be recalculated from time to time during the lifetime of a DP vessel to account for the changes of the vessel's conditions, such as changes of onboard equipment and vessel structures that could affect the environment loads on the vessel or thrust output. DPOs should be fully familiar with the limits given in the DP Capability Plots for the fully intact thruster configuration, for loss of most effective thruster and for worst case failure. The DP footprint plots are also to be produced on board. The footprint plots are actual measurements of the vessel's DP station keeping performance in the actual environmental conditions and thruster configuration at the time the plot was taken. DP footprint plots obtained in harsh environment conditions, can provide a better understanding of the vessel's actual station keeping performance and limitations. The *DP Operation Guidance, Part 2,* provides details on how to generate DP footprint plots. Also, see IMCA M140 regarding DP capability plots.

10.1.4 Position Reference Systems and Sensors

Position reference systems comprise absolute and relative systems. They should be selected with due consideration to operational requirements, both with regard to restrictions caused by the manner of deployment and expected performance in working situations. Vessel owners/operators should consider the advantages of using higher numbers and more reliable position reference systems for the benefit in operational uptime and improved station keeping performance. The *DP Operation Guidance, Part* 2, and the MTS DP Committee TECHOP (D-09-Rev1-Jan-21) PRS and DPCS Handling of PRS (January 2021) provides detailed guidance and recommendations on absolute and relative position reference systems for various operations of DP vessels. Also, refer to IMCA M 252, "*Guidance on position reference systems and sensors for DP operations,*" Rev.1, December 2020.

10.1.5 Recommended DP Control Modes for DP Activities

The DP vessel should be equipped with suitable DP modes and features with due consideration to operational requirements, both with regard to restrictions caused by the activity and performance criteria required to execute the activity safely and successfully. The following control modes are examples relevant to specific DP activities and they are not to be used for other operations without a detailed understanding of the impacts on station keeping and consequences of a loss of position:

- *Target follow:* appliable to riser pull in and lifting, ROV support, shuttler offtake, accommodation support (Note: There are a number of variations on Target Follow and the specificity of the relative position station keeping demands drives the functional requirements).
- *Heavy lift:* applicable to lifting.
- *External force compensation:* applicable to pipelay, riser pull in and lifting.
- *Weathervane:* appliable to shuttle offtake, floating production unit.

10.1.6 Trials and Checklists

A range of trials and checklists should be provided for each DP vessel and implemented as a verification that the vessel's DP system complies with and is operated in accordance with, applicable standards and guidelines. This verification process should confirm the failure modes and their effects on the systems and equipment analysed in the DP FMEA document (to include the Worst-Case Failure) and the vessel's station keeping ability following its Worst-Case Failure. A typical list of trials and checklists for a DP dive support vessel include:

- *DP FMEA proving trials*: A series of tests used to prove the redundancy concept and the expected effects of failure modes found in the FMEA desktop analysis. These tests should be conducted immediately following launching of a new build vessel and following modifications, additions, and upgrades repairs.
- *Endurance trials*: To prove the operation of the DP system under load conditions for at least 4 h without significant alarms of the DP system for new build or following system modification.
- *Annual DP trials*: Series of tests of fault and failure conditions relevant to the DP System. The tests should be designed to prove system redundancy, as defined in the DP FMEA, the elements of Performance, Protection and Detection that the redundancy concept depends upon, system performance and equipment functionality, to validate repairs and preventive maintenance, and test the operation of protection and detection devices and responses so as to demonstrate that the vessel's DP system remains fit for purpose. For DP MODUs, the testing maybe conducted on an incremental basis throughout the year as opportunities arise but needs to be completed within a twelve-month period.
- *DP mobilisation trials*: A series of tests to be carried out at the start of a contract, subject to client requirements, to demonstrate redundancy and functional limitations of the DP system.
- *DP field arrival trials*: A series of checks and tests that confirm satisfactory performance of the DP system and verify the set-up mode of operation and DP functions.
- *DP location set-up checklist*: A series of checks to demonstrate that the vessel is properly set-up for the location, in particular the satisfactory performance of the position reference systems.
- *Pre-dive checklist*: A series of checks performed prior to commencing diving operations. Main and back up communication tests should be included in this process.
- *ECR checklists*: A series of checks and tests that verify that the vessel's set-up and configuration of systems and equipment meet the requirements of Critical Activity Mode (CAM) or Task Appropriate Mode (TAM).
- *500 m checks*: A series of checks and tests performed before entering the 500 m (1,640 feet) zone of an asset in which set-up mode and functions are verified and

confirmed. Approval is then obtained to operate in close proximity to the asset. Main and back up communication tests should be included in this process.
- *Watch status/6 h checklist*: A series of checks and tests performed by the DPOs to verify and confirm the setup of the DP system prior to taking over the DP watch.
- *Post DP incident trials*: Tests performed to verify that the corrective/repair measures taken following a DP incident have properly addressed the causes of the incident and that the vessel's DP system is in a safe and operable condition. ***Note***: *the Tests done post DP incident may include a "stress test" where the equipment is tested at capacity for a defined period (usually no greater than 15–20 min)).* This is usually followed by a "soak" test or endurance test for up to a couple of hours where heading and position changes/box moves are carried out to validate satisfactory performance of the DP system.
- *Post DP modification trials*: A series of checks and tests that are used to determine the effects of modifications and/or additions on the DP system and the vessel's subsequent station keeping performance.

The *DP Operation Guidance, Part 2,* provides detailed guidance notes on the listed trials and checklists for different DP operations.

10.1.7 DP Operations Manuals

A vessel specific DP operations manual should be prepared for each DP vessel. It is the most important operational document. The DP operation guidance provides recommendation for the operations manual including requirements by classes and industry practices. The manual should contain sufficiently detailed instruction and guidance to enable the vessel to be operated safely in DP and safely execute its intended activities. This will include a clear statement on the DP philosophy for the vessel, the organisation, responsibilities and roles of key DP personnel, training, and competency, watchkeeping and manning, vessel technical data and layout, vessel DP capabilities, operating limits, operating modes, the planning and preparation of DP operations, DP operating procedures, emergency procedures, DP incident handling and alert systems and vessel specific trials and checklists that apply uniquely to the vessel. The manual should specifically address operational interfaces between different vendor systems and equipment to confirm that they are configured and operated properly. This should include interfaces between systems and equipment that are not part of the DP system, but which may affect the DP system, such as Emergency Disconnect Systems (EDS) on MODUs, tensioner systems on Pipe lay vessels, etc. The vessel specific manual may also contain generic content, such as company policies, procedures and standing orders. The manual should represent the way the vessel is operated in DP. For complicated power systems and/or thruster configurations, it may be useful provide the operator with a thruster and generator operating strategy (TAGOS) to assist in the decision on what generators and thrusters to use for different circumstances and different equipment availabilities. The *DP*

Operation Guidance, Part 2, includes very details guidance on the development of the vessel specific DP operations manual with four sections (vessel/rig specific):

- Section 1 Management of DP operations (company/rig specific).
- Section 2 DP system description and operation.
- Section 3 DP operations procedures.
- Section 4 Organisation and responsibility.

10.1.8 Activity Operational Planning

In recognising that exposure to risks manifests itself during vessel operations it is recommended that activities performed by DP vessels should be subject to planning and preparation. In planning and preparing for the activities the following should be considered and, where appropriate:

- Configuration for the Critical Activity Mode (CAM) (Sometimes also referred to as the Safest Mode of Operation (SMO)) or, where appropriate, the Task Appropriate Mode (TAM).
- Preparation of the Activity Specific Operating Guidelines (ASOG), including onboard discussion with all relevant stakeholders as part of the pre-project execution/activity.
- Discussion to be included in pre-project execution readiness checklist.
- Capabilities of the vessel, both intact and residual capability, following Worst Case Failure (WCF).
- Limitations imposed by water depth.
- Consequences of a loss of position and/or heading.
- Limitations imposed upon operations by residual capability.
- SIMOPS and marine vessel interaction and consequences arising from change of status (Green to Blue, Yellow or Red).
- The activity being performed and the necessary time to terminate to bring vessel to a safe position upon the onset of failure.
- Communication and Notification Protocols.

Activities should include day to day operations, any specific operation relevant to the design of the vessel, as well as any unique operations the vessel is called upon to perform. Appropriate measures should be in place to clearly identify critical tasks/operational phases of the activity and to confirm that the vessel is set up in Safest Mode of Operation and operating within post WCF capability. Where a decision has been made to operate in a TAM a separate ASOG covering TAM should be produced. Also, refer to IMCA M 220 regarding Guidance on Operational Activity Planning.

10.1.9 Communications

The vessel should be equipped with the appropriate primary and secondary equipment needed to communicate between all parties (stakeholders) whilst carrying out the intended task. Effective internal and external communications is a key tool to manage risk. Communications in this context include voice, visual (lights/displays) and audible means (alarms). Means of communication are not limited to the above but include integrated IT systems using wireless network technology that combine communications with other features, including AIS and DGNSS. Operational specific visual and voice communications should permit pertinent information flows between the key operating points as well as to and from assets and/or other vessels that might be affected by the operation being carried out. These operating points may be on board the vessel as well as on other facilities involved with the activity. Communication protocols should be set up to provide pertinent information regarding intent, current status of planned as well as unexpected events during the execution of the activity. Continuity of communications during foreseeable emergency situations should be taken into account. Communications should be considered when detailing the roles and responsibilities of key personnel during the planning stages for the intended task, ensuring that a common working language and terminology is used at all times.

10.1.10 DP Planned Maintenance

DP vessels should have a structured planned maintenance system that specifically addresses maintenance of the vessel's DP system equipment and support systems. The planned maintenance program should specifically incorporate a focus on the relevant elements of performance, protection, and detection that the redundancy concept depends upon (Vessel Specific). Planned maintenance should address all equipment that has an impact on the vessel's station keeping capabilities. This should include indirect components such as generator circuit breakers, bus tie breakers, etc. Maintenance should include regular cleaning, calibration, and testing of equipment as outlined in manufacturer's recommendations and industry guidelines. Records of planned and unplanned maintenance should be kept in an auditable format, either hard copy or appropriate electronic format. These records should include vendor service records as well as maintenance performed by vessel personnel. These records should be kept on board for the period specified by the vessel owner/operator. A minimum number of required critical DP spares should be maintained on board. The critical spare inventory should be monitored via a formal inventory program that is closely linked to the planned maintenance system. This should assist in getting back to normal operating condition after equipment failure or DP incident.

10.1.11 DP Incidents

DP vessels should be provided with and operate appropriate DP incident reporting, investigation and closing out procedures. This should be in accordance with vessel owner or operators' and, if applicable, clients' processes. Documented records should be kept and be capable of auditing. Where CAM/SMO, TAM and ASOG are used as tools to manage DP operations a suggested approach is that, apart from the exceptions in the notes below, any reactive change of DP status from GREEN to YELLOW or RED should be regarded as a DP incident, and should be reported, recorded and investigated. It is recommended that reactive YELLOW and RED DP incidents are investigated as soon as practicable after the DP incident and, where relevant, trials are carried out as part of the investigation process. In the event of the occurrence of a DP incident relating to the vessel's configuration as described in the DP FMEA and other documents, it is suggested that the FMEA provider is involved in the incident investigation. This will facilitate lessons learnt to be implemented into the DP FMEA and proving trials program. ***Note***: *The* TECHOP G-04 "Conducting Effective and Comprehensive DP Incident Investigations" 2021 *provides valuable guidance on the conduct of DP Incident Investigations.*

10.1.12 Reporting and Record Keeping

The owners or operators of DP vessels should have an effective reporting and record keeping system. There should be a clear line of reporting DP related items onboard the DP vessel and between each department and this should involve key DP personnel. There is also to be a clear line of reporting between the DP vessel and the company's shoreside management. DP related records should be maintained onboard and, where appropriate, at the company's premises. The recommended onboard documents listed at the end of this chapter should be accessible to key DP personnel and to other interested parties. DP vessels, on occasion, carry out activities which may be unique to project requirements. Records of these activities including pertinent information from Hazards and Risk Identification (HIRAs) should be made and kept onboard for future reference, both for training and familiarisation processes as well as for reference in case of similar project requirements in the future.

10.1.13 Competence

DP vessel owners/operators should operate a structured competence assurance program that is applied to all key DP personnel (including shore-based personnel) with special focus on ensuring vessel and task specific competence. Vessel specific competency should, as a minimum, be demonstrated in the following areas:

- Operational modes.
- DP FMEA familiarisation.
- DP operations manual familiarisation.
- Project/activity requirements.
- Contingency plans, modes and drills.
- Comprehension and adherence to the ASOG/WSOG.

DPO certification is only one part of the competency assurance process for DPOs. Such certification is not to be construed as validation of competency. Vessel owners / operators are recommended to refer to further formal training programs described in IMCA M117 "*The Training and Experience of Key DP Personnel*", comprising the following main features for electrical and engineering officers, where appropriate.

- High voltage (HV) training.
- Power management system (PMS) training.
- Fire and gas detection and emergency shutdowns (ESD) training.
- DP control system manufacturer's maintenance training.

IMCA M117 also provides guidance on structured onboard familiarisation programs for key DP personnel, covering the steps to take to achieve, maintain and enhance competency in DP operations. Vessel owners/operators may consider implementing the *IMCA Competence Assurance and Assessment scheme* (IMCA C002 Rev 3, 2020) which provides detailed programs for achieving and assessing the competency of key DP personnel, including vessel masters, DPOs, chief engineers and engineers in charge of a watch.

10.1.14 Manning

Vessels should comply with the statutory requirements for safe manning. Additionally, for DP operations, manning should be in accordance with the following.

10.1.14.1 DP Bridge Crew

When undertaking critical activities in proximity to surface or sub surface structures, two (2) unlimited DPOs on the bridge capable of operating the vessel both in DP and manual control. Unlimited DPO with a minimum of 3-year experience on a vessel engaged in similar operations, at least 6 months of which should have been on the subject or sister vessel. Experience level should be documented and auditable.

10.1.15 Notes

(1) Client's may have specific stipulations on whether the Master could be included or excluded to meet the minimum DP console/Bridge manning requirement.
(2) Critical activities are those activities where the consequences of equipment failure or loss of position are greater than under normal operating circumstances. For example, critical activities on a DP dive support vessel would include those occasions where the Time to Terminate is long, such as when the diver is inside a welding habitat or where the diver's worksite is inside the conductor tubes at a production facility.
(3) It is recognised that, in practice, there could be challenges for vessel owners/operators to meet the recommended level of experience for DPOs. Vessel owners/operators should recognise the associated risk from inexperienced personnel and have plans in place to address them while striving to reach the recommended experience levels.
(4) For offshore supply vessels, IMCA M 182 "*The Safe Operation of Dynamically Positioned Offshore Supply Vessels*" can be used as alternative manning guidance.

10.1.15.1 Engineers

There should be sufficient licensed engineers on board for all expected operations. At least one licensed engineer should be available at all times, should be on watch during critical activities and should have at least 6 months experience on similar equipment and operations. The engineer should be fully cognisant of DP operations, familiar with the vessel's DP FMEA document and the effects of failures of equipment relating to the position keeping of the vessel. In DP 2 or 3 operations, the engineer should be familiar with the general philosophy of redundancy as it relates to split mechanical, electrical and auxiliary systems.

10.1.15.2 Electrician/Electrical Engineer

If required on board, an electrician should have appropriate high voltage training/certification, if applicable to the vessel. As with vessel engineers, the electrician/electrical engineer should have at least 6 months experience on similar equipment and operations. The electrician should be fully cognisant of DP operations, familiar with the vessel's DP FMEA document and the effects of failures of equipment relating to the position keeping of the vessel.

10.2 Recommended DP Documentation

It is recommended that DP vessel owners/operators should maintain the documentation listed below. The documentation version should be current and kept on board, in addition, where feasible, at the shore-based centres of technical management. Documents that have been superseded should be clearly marked and kept separate from current versions.

- DP System FMEA or FMECA. To be kept up to date, incorporating all modifications and additions since original study, if not in the document itself, then by other traceable means. All records to be kept on board.
- DP FMEA Proving Trials. To be conducted to prove initial DP FMEA and at other times to prove modifications and additions to the DP system. DP FMEA Proving Trials should be repeated every five years. Findings and recommendations to be addressed in accordance with their criticality. All records to be kept on board.
- *Annual DP trials.* To be conducted annually. Findings and recommendations should be addressed in accordance with their criticality. Previous trials reports and associated close out documentation to be kept on board.
- *DP capability plots.* Hard copy DP capability plots relevant to the vessel's areas of operations to be readily accessible to DPOs at the DP control location.
- *DP footprint plots.* Hard copy DP footprint plots to be taken by DPOs and kept on board.
- *Service reports concerning the DP system.* Complete history of service reports to be kept on board.
- *Details of all DP related modifications and additions.* Records of all DP related modifications and additions to be kept on board complete with interface and testing information.
- *Vessel audit reports and DP audits and inspection reports.* Complete history of all audit reports, DP audits and inspection reports, incident findings and close outs to be kept on board.
- *DP operations manual.* Vessel specific DP operations manual, to be readily accessible at the DP control location and used by the DPOs as a reference for conducting DP operations.
- *DP incident reports.* Records of all DP station keeping and other DP related incidents to be kept on board, incident investigation records and close outs.
- *DP location and watchkeeping checklists (bridge and engine room).* Records of all DP location and watchkeeping checklists to be kept on board for the period set by the vessel owner/operator and, where relating to a DP incident, permanently stored in retrievable archives.
- *DP related drills and emergency response drills.* Records of DP related drills and emergency response drills to be kept on board in retrievable archives.
- *DP fault log.* Records of all faults related to the DP system to be kept on board permanently in retrievable archives.

- *DP data logging.* Where the vessel has DP data logging facilities, electronic records should be kept on board for the period set by the vessel owner/operator and, where relating to a DP incident, permanently stored in retrievable archives.
- *DP alarm printer readouts.* Hard copy records of the DP alarm printer readout to be kept on board for the period set by the vessel owner/operator and, where relating to a DP incident, permanently stored in retrievable archives.
- *DP familiarisation and competency records.* All records relating to vessel specific DP familiarisation and competency for DPOs, engineers and electricians to be kept on board permanently in retrievable archives.
- *Résumés of all key DP personnel, copies of certification and qualifications, records of DP watchkeeping hours to be maintained on board.* Original DPO certificates and DP logbooks to be held by the DPOs onboard the vessel.

Annex 1
Available Thrust Assessment

This appendix provides guidelines for the determination of the thrust generated by various types of thrusters. It also addresses the interactions of thrusters (thruster-thruster, thruster-hull, thruster-current, which often result in a reduction of the available thrust. The available thrust from this appendix may be used for preliminary studies for the design of the DP system. Manufacturer's test data of full scale or suitable model test for the thrust output of thrusters are to be used whenever possible for further verification.

A1.1 Thrust at Bollard Pull

The achievable bollard pull for a thruster is basic data for the DP system performance assessment. The graphs below indicate the available bollard pull for typical thruster configurations (including conventional main propulsion arrangements), for propellers with nozzles and open propellers. The propeller disk area load is the ratio of the thrust per motor input power in kN/kW and the motor power per propeller disk area $A = D^2\pi/4$, in m^2. For open propellers, the following equation can be used to calculate the available bollard pull thrust (the units of measure are in SI (MKS and US) systems, respectively) (Fig. A1.1):

$$T_0 = K \cdot (P \cdot D)^{2/3}$$

where

T_0 bollard pull, in N (kgf, lbf)
P propeller power, in kW (PS, hp)
D propeller diameter, in m (m, ft)
K 848 (70.4, 70.3)

F. Karkori, *Dynamic Positioning Systems*, Springer Series on Naval Architecture, Marine Engineering, Shipbuilding and Shipping 21,
https://doi.org/10.1007/978-3-031-59173-0

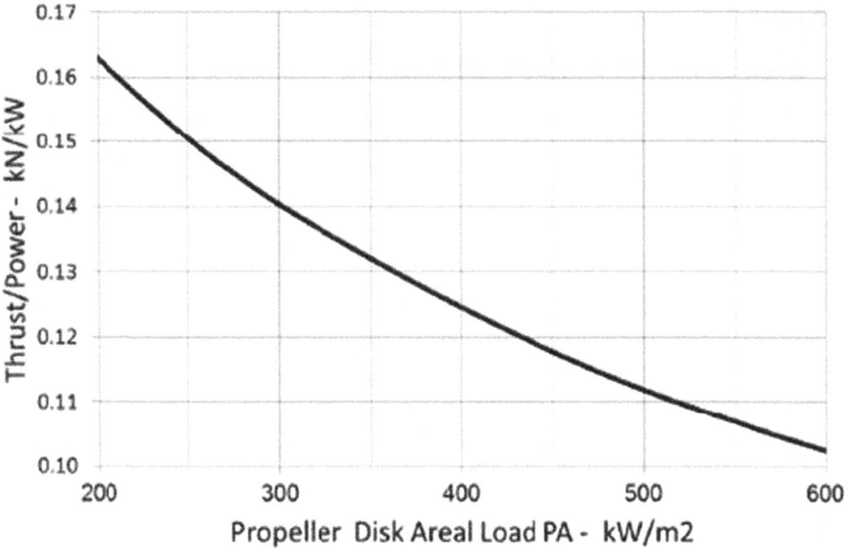

Fig. A1.1 Open propellers

For ducted propellers, the following equation can be used to calculate the available bollard pull thrust (Fig. A1.2):

$$T_0 = K \cdot (P \cdot D)^{2/3}$$

where

T_0 bollard pull, in N (kgf, lbf)
P propeller power, in kW (PS, hp)
D propeller diameter, in m (m, ft)
K 1,250 (103.8, 103.7)

A1.2 Thruster-Thruster Interaction

The reduction of the thrust output due to thruster-thruster interaction may depend on the following:

(1) Distance between the thrusters.
(2) Azimuth of the thrusters.
(3) Diameter of the thruster propeller.
(4) Thruster load.
(5) Thruster design/configuration (i.e., degree of tilt of the propeller and/or nozzle axis).

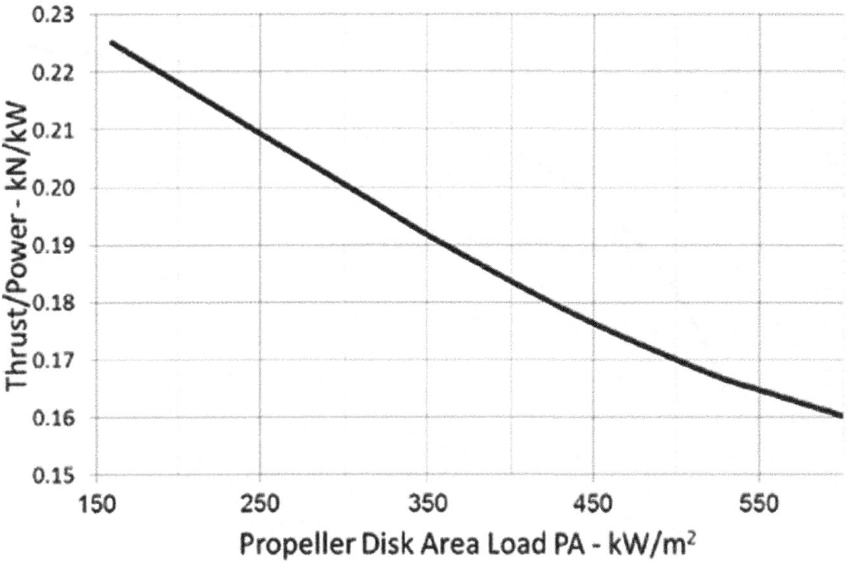

Fig. A1.2 Ducted propellers

The following paragraphs describe the thrust reductions for two principal identical thruster configurations.

A1.2.1 In Line Tandem Condition

Figure A1.3 depicts thrusters in line tandem configuration. The rear thruster operates directly downstream of the other thruster in open water.

The thrust reduction ratio t defined below for the downstream thruster can be calculated as follows (Fig. A1.4):

$$t = \frac{T}{T_0} = 1 - 0.75^{\left(\frac{x}{D}\right)^{\frac{2}{3}}}$$

Fig. A1.3 Thrusters configuration in tandem condition

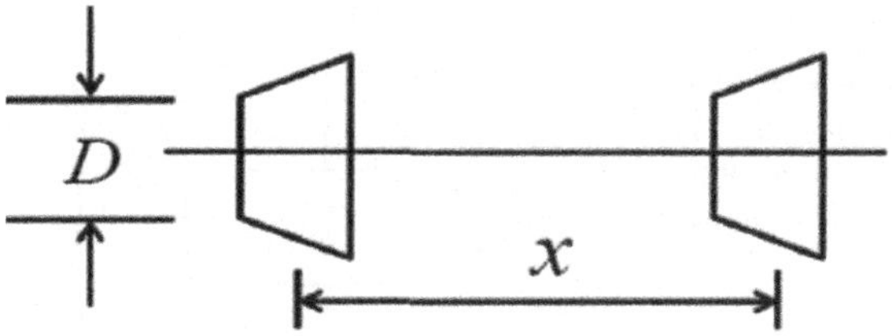

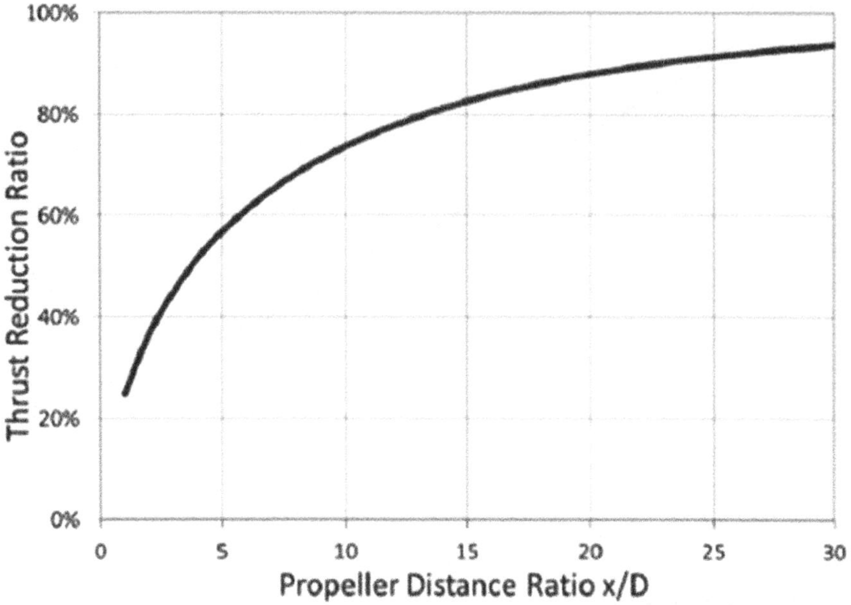

Fig. A1.4 Thrust reduction of the downstream thruster in open water

where

- x distance between the two thrusters, in m (m, ft)
- D thruster diameter, in m (m, ft)
- T_0 bollard pull thrust in open water, in N (kgf, lbf)
- T thrust of the downstream thruster, in N (kgf, lbf)
- t thrust reduction ratio

A1.2.2 Upstream Thruster Turned Tandem Configuration

Figure A1.5 depicts the thruster in upstream turned tandem configuration.

The thrust reduction ratio t defined below for the downstream thruster considering steering angles of the upstream thruster can be calculated as follows (Fig. A1.6):

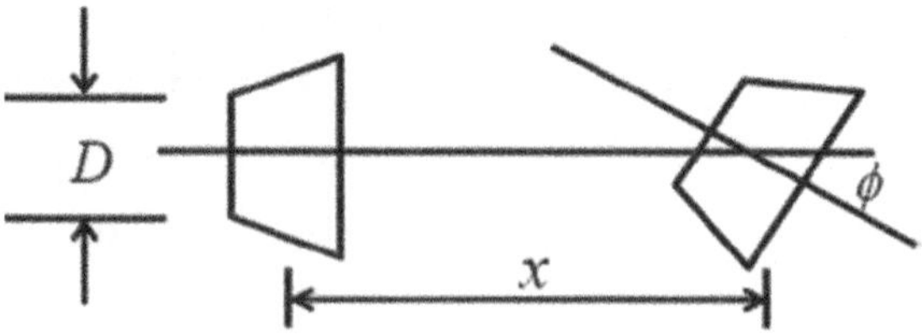

Fig. A1.5 Thrusters configuration in tandem condition turning the upstream thruster

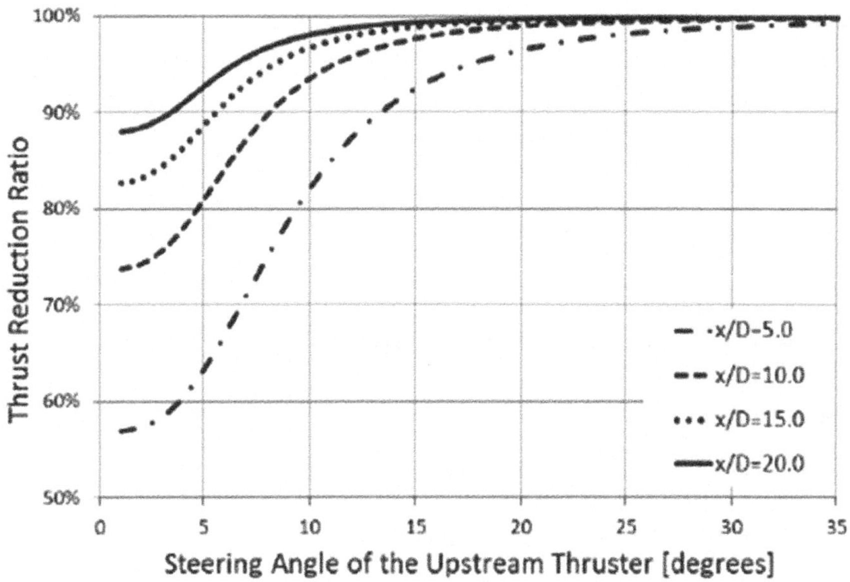

Fig. A1.6 Thrust reduction of the downstream thruster considering steering angles of the upstream thruster

$$t\phi = t + 1 - t$$

where

ϕ steering angle, in degrees
t thrust reduction ratio at zero steering angle
$t\phi$ thrust reduction ratio at steering angle, ϕ

A1.2.3 Forbidden Zones

Forbidden zones, sometimes called barred sectors, may be utilised in thruster control to avoid excessive loss due to thruster-thruster interactions. That can be achieved by limiting certain orientations of azimuth thrusters. The forbidden zones can be calculated using a simple algorithm based on the thruster-thruster interaction effect presented above. The range of the zones shown in Table A1.1 depends on the distance between the thrusters and their diameters, and it could be determined by the following method:

Table A1.1 Range of forbidden zone for different x/D

x/D	Angle (degrees)	x/D	Angle (degrees)	x/D	Angle (degrees)
1	30	6	17.8	11	14.2
2	26.3	7	16.8	12	13.8
3	22.8	8	16	13	13.3
4	20.6	9	15.3	14	13
5	19	10	14.7	15	12.6

$$\phi_f \text{ is the angle which minimises the value of } \frac{T_d}{t\phi \cdot \cos\phi} (0^\circ < \phi < 45^\circ)$$

where

ϕ_f range of forbidden zone, in degrees
T_d demanded thrust, in N (kgf, lbf)
$t\phi$ thruster reduction ratio at steering angle ϕ

Figure A1.7 and Table A1.1, x is the distance between the two thrusters and D is thruster diameter.

A1.3 Thruster-Hull Interaction

This section provides the methods for the calculation of thrust degradation due to thruster-hull interaction. Consideration of thruster tilt is also provided.

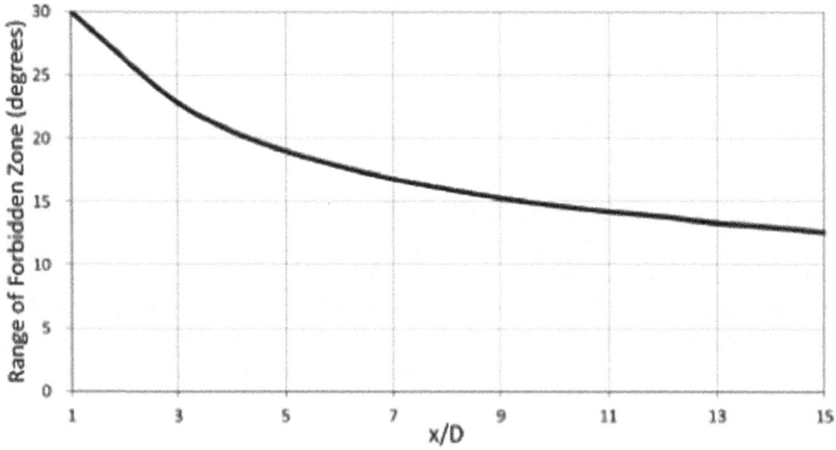

Fig. A1.7 Range of forbidden zone

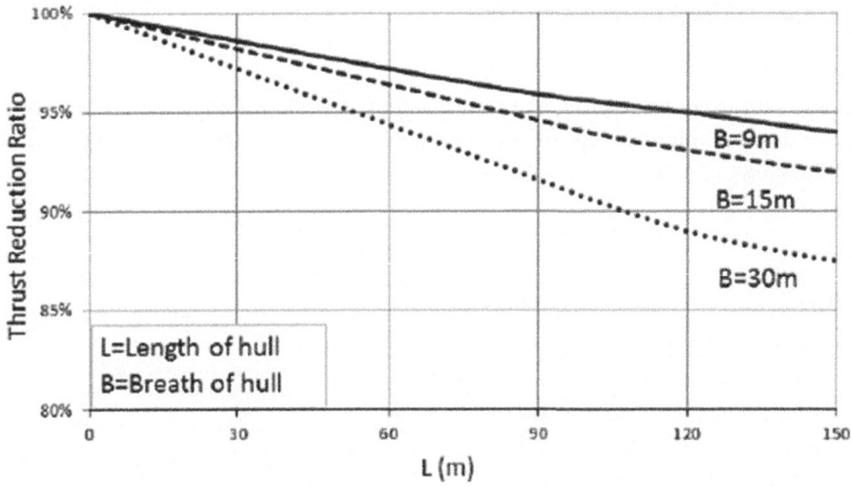

Fig. A1.8 Thrust reduction ratio due to hull friction

A1.3.1 Friction

Thrust degradation due to hull friction is related to the length and breadth of the downstream flow along the hull. The graph below can be used for the assessment of the thrust reduction ratio tf due to the hull friction (Fig. A1.8).

A1.3.2 Coanda Effect

The Coanda effect is related to the bilge radius and the length of the flow underneath the hull. If no detailed data available, the thrust reduction ratio tc due to Coanda effect can be taken as 97%.

A1.3.3 Pontoon Blockage

The blockage of the downstream flow due to presence of the pontoon occurs when a downstream is orientated towards the pontoon, such as the downstream of a thruster on one pontoon is directed towards the opposite pontoon. The reduction of the thrust output due to the pontoon blockage can be calculated using the formula below:

$$t_p = 0.8K \cdot \left(\frac{LP}{55}\right)$$

A1.3.4 Tilted Thruster/Nozzle

A tilted thruster can improve the thrust output. For the tilt angle range from 0 to 8 degrees, the following equation can be applied to determine the improvement of thrust reduction ratio:

$$t_h = t_f \cdot t_c \cdot t_p + \left(1 - t_f \cdot t_c \cdot t_p\right) \cdot C$$

where

t_h total thrust reduction ratio of thruster-hull interaction
t_f thrust reduction ratio due to friction
t_c thrust reduction ratio due to Coanda effect
t_p thrust reduction ratio due to pontoon blockage
C tilt thruster correction factor

Figure A1.9 Correction factor, C, as the function of the tilt angle of the propeller shaft.

A1.3.5 Thruster-Current Interaction

Current inflow may reduce thrust output of the thrusters and the reduction of the thrust can be calculated using the graphs or equations given below for ducted or open propellers at current speeds between 0 and 2 m/s.

For ducted propellers with current speed ranges from 0 to 2 m/s, the following equation can be applied (Fig. A1.10):

$$EFC = 1 - \left(K_1 / PA\right)^{0.25} \cdot K_2 \cdot V_C$$

where

EFC thrust reduction ratio due to current
V_c current speed, in m/s (m/s, ft/s)
PA propeller disk area load, in kW/m^2 (PS/m^2, hp/ft^2)
K_1 400 (544, 50)
K_2 0.11 (0.11, 0.034)

For open propellers with current speed in the range from 0 to 2 m/s, the following equation can be applied:

$$EFC = 1 - (K_1/PA)^{0.28} \cdot K_2 \cdot V_c$$

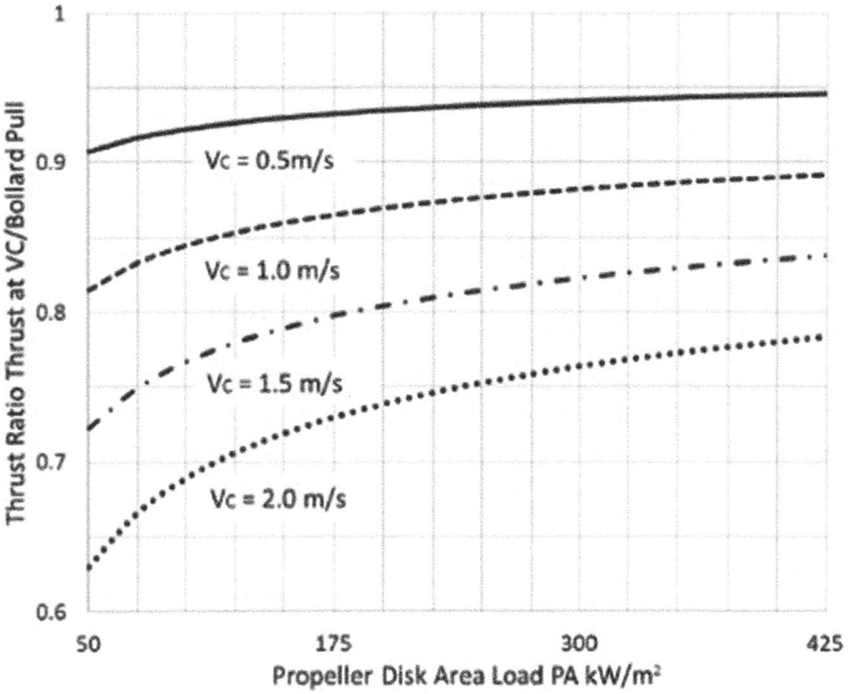

Fig. A1.9 Thrust ratio for ducted propellers

where

EFC	thrust reduction ratio due to current
V_c	current speed, in m/s (m/s, ft/s)
PA	propeller disk area load, in kW/m^2 (PS/m^2, hp/ft^2)
K_1	10 (136, 1.25)
K_2	0.11 (0.11, 0.034)

A1.4 Tunnel Thrusters

The available thrust generated by a tunnel thruster is highly affected by the tunnel installation, location, and the geometry and configuration of the hull. The assessment of the tunnel thruster performance is to be based on axial flow pump hydrodynamics, not on marine propeller theories. If no data are available, the available thrust of a tunnel thruster can be estimated according to the following procedure which is based on axial flow pump theory adapted to the specific conditions of tunnel thrusters (Fig. A1.11).

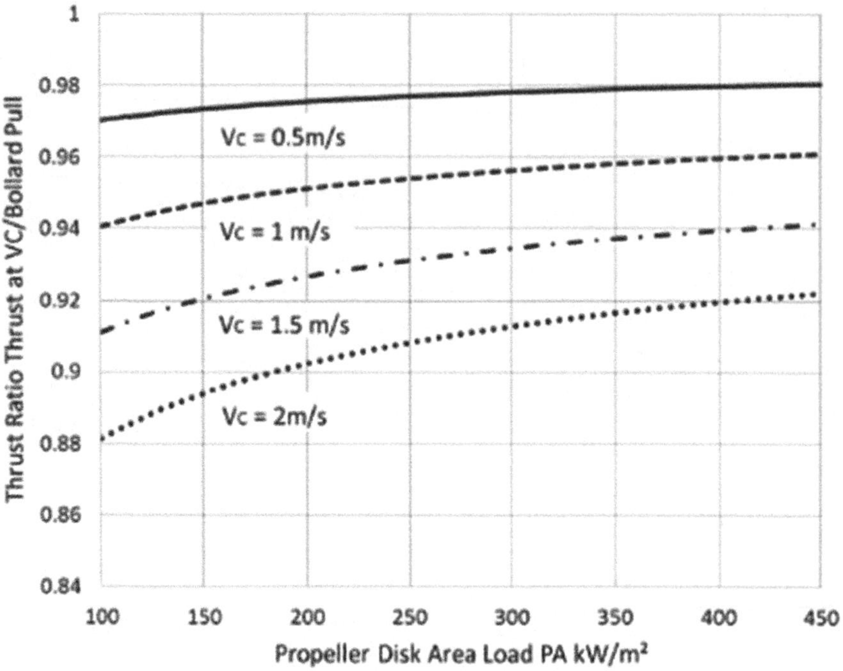

Fig. A1.10 Thrust ratio for open propellers

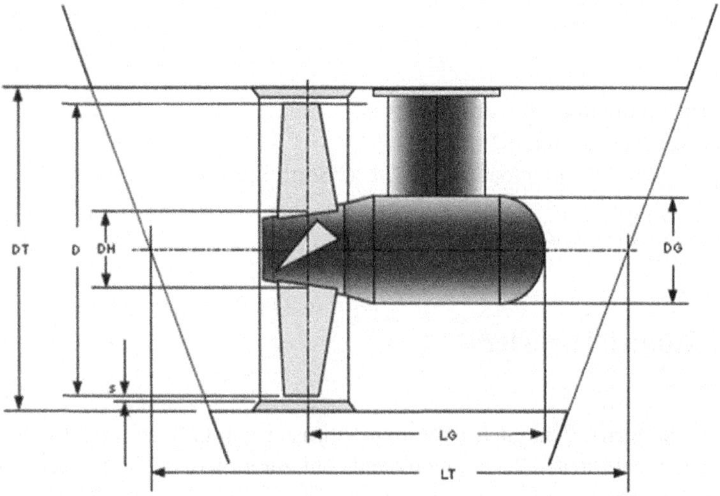

Fig. A1.11 Dimensions of transverse tunnel thrusters

A1.4.1 Simplified Method

At the preliminary design phase of a project, a simplified equation, provided below, for the calculation of available thrust can be used, which is based on typical average tunnel thruster configurations.

$$T = \sqrt[3]{\rho \cdot \pi \cdot \left(K \cdot P \cdot \eta_p \cdot \frac{DT}{TL}\right)^2}$$

where

T	thrust, in N (kgf, lbf)
ρ	mass density of seawater = 1025 kg/m^3 SI and MKS units = 63.9 lb/ft^{3}US units
K	1000 (24, 97.2)
P	power at the tunnel thruster propeller, in kW (PS, hp) = $P_{mot} \cdot \eta_{TG}$
P_{mot}	motor output power, in kW (PS, hp)
η_{TG}	efficiency of the thruster gear
η_p	efficiency of the propeller (impeller) = 0.8
DT	tunnel diameter, in m (m, ft)
TL	1.65 for tunnels of average length, no grids, hull angle δ = 70 degrees, and conical tunnel entries

A1.4.2 Detailed Method

When more information is available at the later design stage for tunnel thrusters, the detailed method described below can be used (Fig. A1.12).

A1.4.2.1 Thrust Loss Factor

(1) Hydrodynamic loss caused by the tunnel entry/exit geometry

E = 0.1 for rounded entry/exit shape
E = 0.25 for conical entry shape
E = 0.5 for sharp entry

(2) Venturi effect loss factor

$V = 0.15$ (typical value for tunnel thruster)

(3) Friction loss factor

$$R = CF \cdot A \cdot LT/DT$$

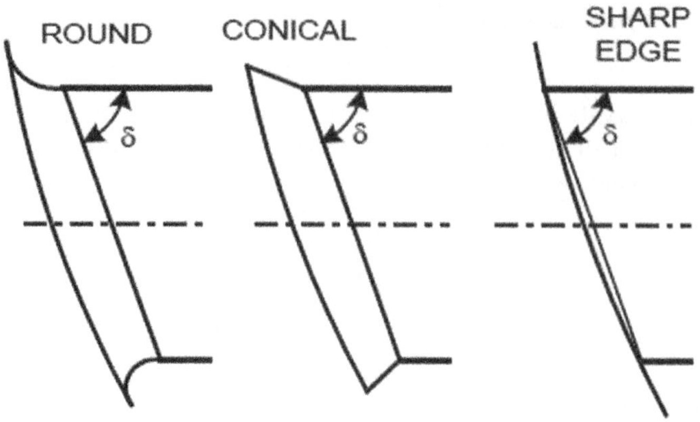

Fig. A1.12 Tunnel configuration

where

CF coefficient of friction assuming exposed weld seams and minor marine growth on surfaces = 0.05

A factor for the increased velocity inside the tunnel due to the thruster gear housing $= 1 + LG/LT\{\frac{1}{\left[1-\left(\frac{DG}{DT_1}\right)^2\right]^2(1-\frac{DG}{DT}{}_1)} - 1\}$

LG, LT, DG, and DT are as given in Fig. A1.11.

(4) Loss due to protective bars

This applies to tunnels equipped with protective bars (grids) at the openings.

$$G = K \cdot SF \cdot [AG/(DT^2 \cdot \frac{\pi}{4} - AG]^{\frac{4}{3}}$$

where

SF grid shape factor (refer to Fig. A1.13

AG projected grid area in m^2 (m^2, ft^2)
DT tunnel diameter in m (m, ft)

(5) Loss due to hull inclination

$$H = 0.3 \cdot \cos(\delta) + 0.2 \cdot \cos 2(\delta)$$

where

H influence of the hull angle
δ in degrees (shown in Fig. A1.13)

Fig. A1.13 Protective bars cross section

(6) Total loss factor

$$TL = 1 + E + V + R + G + H$$

A1.4.2.2 Available Thrust

For tunnel thrusters, the following equations can be used to calculate the available thrust:

$$T = \sqrt[3]{\rho \cdot \pi \cdot \left(K \cdot P \cdot \eta_p \cdot \frac{DT}{TL}\right)^2}$$

where

T	thrust in N (kgf, lbf)
ρ	mass density of seawater = 1,025 kg/m^3 [SI and MKS units] = 63.9 lb/ft3 [US units]
P	power at the tunnel thruster propeller, in kW (PS, hp) = $P_{mot} \cdot \eta_{TG}$
P_{mot}	motor output power, in kW (PS, hp)
η_{TG}	efficiency of the thruster gear = 0.95
η_P	efficiency of the propeller (impeller) = 0.8
DT	tunnel diameter, in m (m, ft)
TL	total loss factor
K	1,000 (24, 97.2)

Annex 2
Rudder Forces

For vessels with conventional propulsion arrangements, the propeller jet and its interaction with the rudder can be used to develop transverse forces in the stern of the vessel in lieu of thrusters during dynamic positioning operations. The following graphs can be used for the estimate of achievable rudder forces in transverse and longitudinal (ahead) direction for rudder with different nozzle configurations. Manufacturer's test data of full scale or suitable model test for the rudder forces are to be used whenever possible for further verification (Figs. A2.1, A2.2, A2.3 and A2.4).

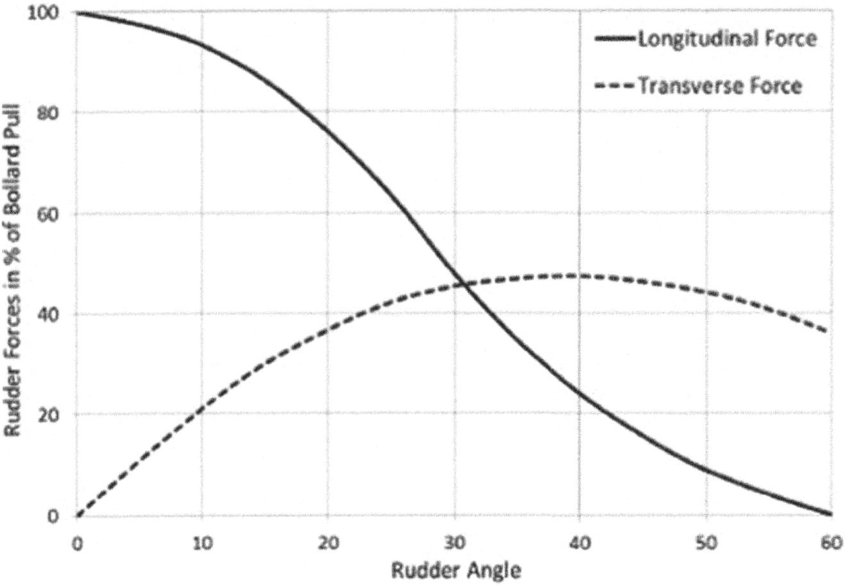

Fig. A2.1 High-performance rudder

F. Karkori, *Dynamic Positioning Systems*, Springer Series on Naval Architecture, Marine Engineering, Shipbuilding and Shipping 21,
https://doi.org/10.1007/978-3-031-59173-0

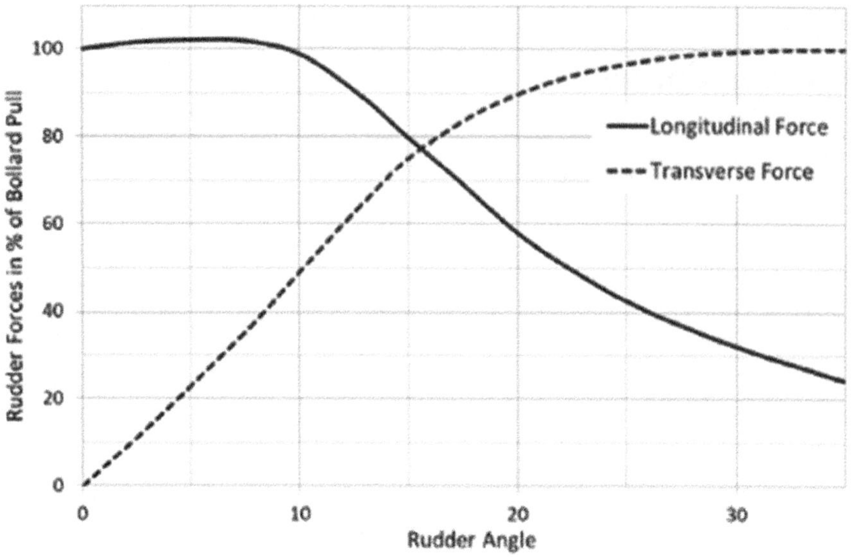

Fig. A2.2 Steerable nozzle with flap

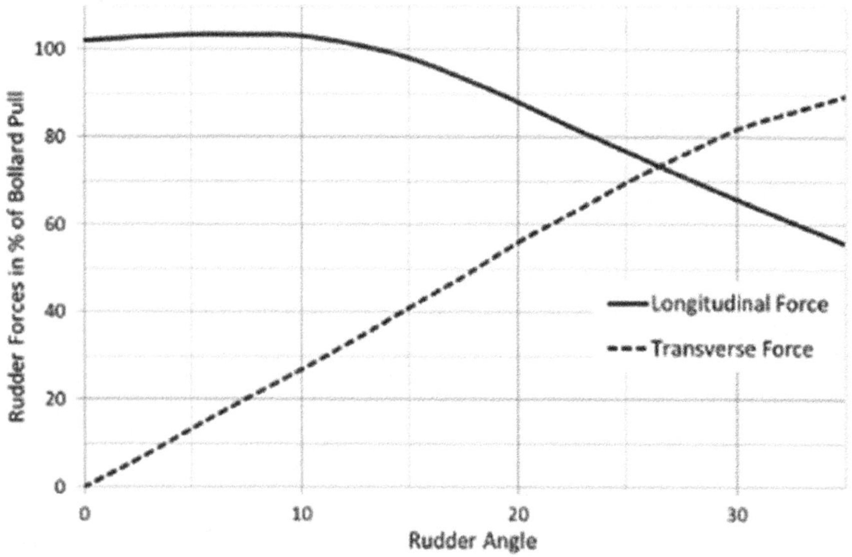

Fig. A2.3 Steerable nozzle with fixed fin

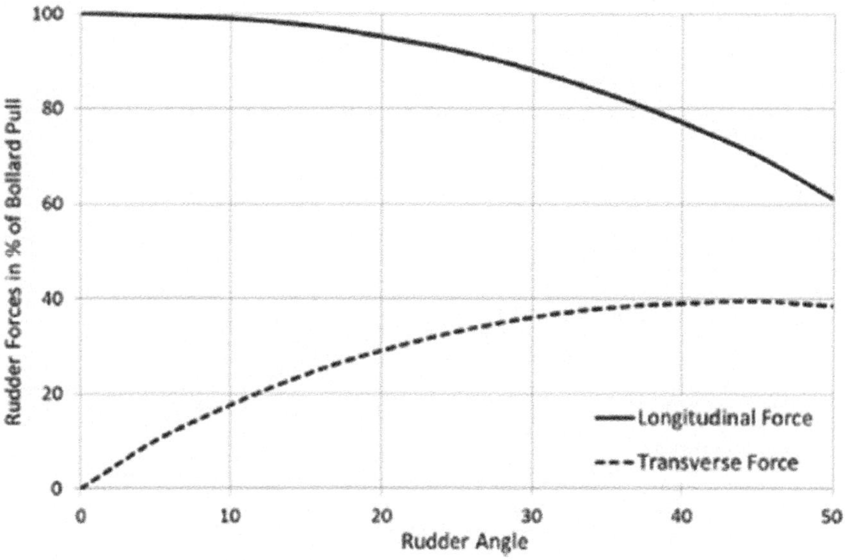

Fig. A2.4 Conventional rudder

Bibliography

1. API RP 2SK: Recommended practice for design and analysis of station keeping systems for floating structures
2. IEC 60812: Analysis techniques for system reliability—procedure for Failure Mode and Effects Analysis (FMEA)
3. IMCA M 04/04: Methods of establishing safety and reliability of dynamic positioning systems
4. IMCA M 103: Guidelines for design and operation of dynamically positioned vessels
5. IMCA M 166: Guidance on Failure Mode and Effects Analysis (FMEA)
6. IMCA M 206: A guide to electrical power and control systems
7. IMCA M 247: Guidance to identify DP system components and their failure modes
8. IMCA: Specification for DP capability plots
9. IMO 2000 HSC Code: International code of safety for high-speed craft as amended
10. IMO 2009 MODU Code: Code for the constriction and equipment of mobile offshore drilling units, as amended
11. IMO MSC.1 Circular 1580: Guidelines for vessels and units with dynamic positioning (DP) systems
12. MTS: DP operation guidance
13. MTS: DP vessel design philosophy guidelines
14. MTS: Technical and operational guidance (TECHOP) TECHOP_D-01-Rev1-Jan21 addressing C3EL2 to eliminate Single Point Failures (C3EL2-cross-connections, commonality, external interfaces and influences)
15. MTS: Technical and operational guidance (TECHOP) TECHOP_D-07-Rev1-Jan21, a method for the fault ride-through capability of DP vessels with hv power plant
16. OCIMF: Dynamic positioning failure mode effects analysis assurance framework risk-based guidance
17. OCIMF: Prediction of wind and current load on VLCCs

F. Karkori, *Dynamic Positioning Systems*, Springer Series on Naval Architecture, Marine Engineering, Shipbuilding and Shipping 21,
https://doi.org/10.1007/978-3-031-59173-0